Lecture Notes in Economics and Mathematical Systems

374

W0051078

Georg Pflug   Ulrich Dieter (Eds.)

Simulation and Optimization

Proceedings of the International Workshop on Computationally Intensive Methods in Simulation and Optimization held at the International Institute for Applied Systems Analysis (IIASA), Laxenburg, Austria, August 23-25, 1990.

Springer-Verlag

Berlin Heidelberg New York
London Paris Tokyo
Hong Kong Barcelona
Budapest

Editors

Prof. Dr. Georg Pflug
Institut für Statistik und Informatik
Universität Wien
Universitätsstraße 5
A-1010 Wien, Austria

Prof. Dr. Ulrich Dieter
Institut für Statistik
Technische Universität Graz
Lessingstraße 27
A-8010 Graz, Austria

ISBN-13: 978-3-540-54980-2   e-ISBN-13: 978-3-642-48914-3
DOI: 10.1007/978-3-642-48914-3

Typesetting: Camera ready by author

42/3140-543210 - Printed on acid-free paper

Simulation and Optimization:
Proceedings of the International Workshop on Computationally Intensive Methods in Simulation and Optimization

Contents

Part II: Optimization and Stochastic Optimization

Part III: Random Numbers

Foreword

This volume contains selected papers presented at the "International Workshop on Computationally Intensive Methods in Simulation and Optimization" held from 23th to 25th August 1990 at the International Institute for Applied Systems Analysis (IIASA) in Laxenburg, Austria. The purpose of this workshop was to evaluate and to compare recently developed methods dealing with optimization in uncertain environments. It is one of the IIASA's activities to study optimal decisions for uncertain systems and to make the result usable in economic, financial, ecological and resource planning. Over 40 participants from 12 different countries contributed to the success of the workshop, 12 papers were selected for this volume.

Prof. A. Kurzhanskii
Chairman of the Systems and
Decision Sciences Program
IIASA

Preface

Optimization in an random environment has become an important branch of Applied Mathematics and Operations Research. It deals with optimal decisions when only incomplete information of the future is available. Consider the following example: you have to make the decision about the amount of production although the future demand is unknown. If the size of the demand can be described by a probability distribution, the problem is called a *stochastic optimization problem.*

Any solution method of such a problem contains a certain interface between simulation and optimization which is crucial for the accuracy and efficiency of its solution. Two main approaches of organizing the interface have been investigated in greater details:

- *First simulation, then optimization*: The random variables appearing in the model are replaced by a computer generated random variates. This means that the probability measures are either known or approximated by empirical discrete distributions. In both cases efficient sampling methods are needed. For each realization of the data the resulting large scale program is solved and taken as an approximate solution of the original problem. In this second part efficient procedures for large scale (deterministic) programs are necessary as well.

- *Simulation interleaved with optimization*: This method is organized in iterative cycles. At each cycle just one step of an optimization algorithm is performed and after that a new random observation is added. The best known examples of such algorithms were introduced by Robbins/Monro and Kiefer/Wolfowitz type and further developed by other authors. Here a gradient or quasi gradient step is performed in a direction given by a random observation. Other examples of this group of algorithms are the *adaptive experimental design methods*, where the shape of the response function is locally approximated and optimal points for the next simulation experiments are determined.

For an evaluation of the strength and the applicability of the two methods one has to consider the relationship between the decision making and the source of randomness and to distinguish two cases:

- *The probability distributions do not depend on the decision making.* Many problems in economic, financial and environmental planning are of this kind. Here the unknown random quantities are future demand, future interest rate, stock prices or weather conditions. In all these cases both groups of methods are applicable.

- *The decision influences the probability distributions.* Production or communication systems are modelled by stochastic systems and their performance is measured in terms of the steady–state behaviour. Any decision on parameters of the system changes the steady state. For this class of problems only the *simulation interleaved with optimization* approach is applicable, since the probability law for sampling is not known in advance. Here an important task is to determine gradients or quasi-gradients of the performance function which have small variance.

The first part of this volume contains papers which use the *simulation interleaved with optimization* approach. In the first three papers gradient estimates for the performance of discrete event dynamic systems are studied. The fourth paper deals with regression and design methods. Part II contains a collection of papers on stochastic and deterministic optimization. Part III is devoted to random number generation, a field which is crucial for stochastic optimization.

The editors wish to express their gratitude to the sponsors of the workshop, the Austrian Ministry of Science and Research and to IIASA Laxenburg. Furthermore, they thank Springer Verlag for including the proceedings in the Lecture Notes Series.

G. Pflug, Wien   U. Dieter, Graz

Performance evaluation for the score function method in sensitivity analysis and stochastic optimization

Søren Asmussen

Chalmers University of Technology, Göteborg, Sweden*

Reuven Rubinstein

Technion, Haifa, Israel

Abstract

Estimating systems performance $\ell(\rho) = \mathbb{E}_\rho L$ and the associated sensitivity (the gradient $\nabla \ell(\rho)$) for several scenarios via simulation generally requires a separate simulation for each scenario. The score function (SF) method handles this problem by using a single simulation run, but little is known about how the estimators perform. Here we discuss the efficiency of the SF estimators in the setting of simple queueing models. In particular we consider heavy traffic (diffusion) approximations for the sensitivity and the variances of the associated simulation estimators, and discuss how to choose a 'good' reference system (if any) in order to obtain reasonably good SF estimators.

1 Introduction

Let $\{L_t\}_{t=1,2,\dots}$ be a discrete time regenerative process with cycle length C and distribution $\mathbb{P}_\rho(\cdot)$ depending on a parameter ρ. Assuming that C has an aperiodic distribution and that $\mathbb{E}_\rho C < \infty$, it is well known that the expected steady-state performance can be represented as

$$\ell(\rho) = \mathbb{E}_\rho L = \frac{\mathbb{E}_\rho R}{\mathbb{E}_\rho C} = \frac{r(\rho)}{c(\rho)}, \tag{1.1}$$

where $R = \sum_1^C L_t$, and similarly when L_t is a continuous-time regenerative process.

Usually, $\ell(\rho)$ is not available analytically, and one may want to estimate $\ell(\rho)$ as well as the associated sensitivities $\nabla^p \ell(\rho)$, $p = 1, 2, \dots$ by using the Monte Carlo method. One established approach is the *score function* (SF) *method* or *likelihood ratio* (LR) *method*, see e.g. [3], [6], [7], [8], [4] which has the particular advantage of allowing to perform the estimation *simultaneously*. That is, one can use a single simulation run (with reference parameter say ρ_0) for several scenarios $\rho_1, \rho_2, \dots$ instead of, as is generally required, a separate simulation for each scenario.

The purpose of this paper is to survey some recent results of the authors ([3], [4]) on the performance and the efficiency of the SF simulation estimators. In particular we consider the following problems:

Present address Institute of Electronic Systems, Aalborg University, Fr. Bajersvej 7, DK–9220 Aalborg Ø, Denmark

(Section 2) What is the analogue of the standard diffusion approximation set–up in heavy traffic when we consider not only the steady state mean $\ell(\rho)$ and $\{L_t\}$ but also the sensitivity $\nabla\ell(\rho)$ and the SF process?

(Section 3) Which of the various variants of the SF method estimators is the more efficient? What is the heavy traffic performance?

(Section 4) Given a stable queueing model with traffic intensity $\rho < 1$, what is a 'good' reference system (if any) to simulate in order to obtain reasonable good SF estimators? What is the optimal reference traffic intensity ρ_0 of such a system? For a given reference value ρ_0, how does the SF estimators perform in various ranges of ρ?

For the sake of simplicity, we shall use the reference parameter $\rho_0 = \rho$ in the study of the sensitivity $\nabla\ell(\rho)$ (Sections 2–3) and consider only the steady state mean $\ell(\rho)$ in Section 4 where we turn to the study of the case $\rho_0 \neq \rho$.

2 Heavy traffic properties of the SF process

It is assumed throughout the paper that $\{L_t\}_{t=1,2,\ldots}$ is driven by an input sequence $\{Y_t\}_{t=1,2,\ldots}$, and that C is a stopping time, that is, the occurence of the event $\{C = t\}$ is determined by $Y_1, \ldots, Y_t$ alone. We also assume that the input sequence is i.i.d. and let $f_\rho(y)$ be the common density of the Y_t. For a typical example, one may think of $\{L_t\}$ as the waiting time process in the GI/G/1 queue.

The score function is

$$\tilde{S}_t = \sum_{i=1}^{t} \nabla \log f_\rho(Y_i) = \sum_{i=1}^{t} \frac{\nabla f_\rho(Y_i)}{f_\rho(Y_i)}.$$

It is related to the sensitivity as follows. Differentiating (1.1), we get

$$\nabla\ell(\rho) = \nabla\frac{r(\rho)}{c(\rho)} = \frac{\nabla r(\rho)}{c(\rho)} - \frac{r(\rho)\nabla c(\rho)}{c(\rho)^2} \tag{2.1}$$

Here two alternative expressions occur in the literature. Consider for the sake of simplicity first $\nabla r(\rho)$. Then, in the terminology of [3], the *straightforward* (or *crude*) expression is

$$\nabla r(\rho) = \mathbb{E}_\rho \tilde{S}_C \sum_{t=1}^{C} L_t, \tag{2.2}$$

whereas the *sophisticated* (or *efficient*) one is

$$\nabla r(\rho) = \mathbb{E}_\rho \sum_{t=1}^{C} L_t \tilde{S}_t. \tag{2.3}$$

It will be argued in Section 3 that (2.3) is the preferable version. It leads to

$$\nabla\ell(\rho) = \frac{\mathbb{E}_\rho \sum_{t=1}^{C} L_t S_t}{\mathbb{E}_\rho C} - \frac{\mathbb{E}_\rho \sum_{t=1}^{C} L_t}{\mathbb{E}_\rho C} \frac{\mathbb{E}_\rho \sum_{t=1}^{C} S_t}{\mathbb{E}_\rho C} \tag{2.4}$$

$$= \mathbb{E}_\rho LS - \mathbb{E}_\rho L \mathbb{E}_\rho S = \mathbf{Cov}_\rho(L, S), \tag{2.5}$$

where $\{S_t\}$ is the SF process, i.e. the process that we obtain from the scores $\{\tilde{S}_t\}$ by 'resetting at regeneration points', i.e. letting

$$
\begin{aligned}
S_t &= S_t \\
&= \nabla \log f_\rho(Y_1) + \ldots + \nabla \log f_\rho(Y_t), \quad 1 \le t \le C, \\
S_t &= \nabla \log f_\rho(Y_{C+1}) + \ldots + \nabla \log f_\rho(Y_t), \quad C+1 \le t \le C^{(1)},
\end{aligned}
$$

(where $C^{(1)}$ is the second regeneration point) and so on. Noting that the processes $\{S_t\}$ and $\{(L_t, S_t)\}$ are regenerative, this provides a representation of the gradient $\nabla \ell(\rho)$ in the regenerative framework. A typical sample path of the bivariate process $\{(L_t, S_t)\}$ is given on Figure 1.

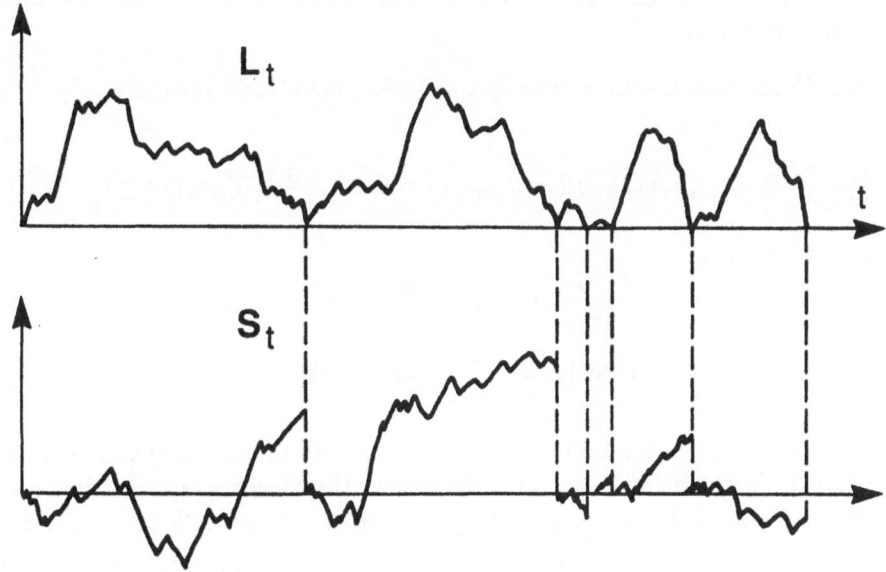

Figure 1:

For a great number of queueing models, the heavy–traffic behaviour of $\{L_t\}$ alone can be described by means of reflected Brownian motion (RBM) $\{\bar{\xi}_t\}$ (in its standard version, with drift -1 and variance 1). More precisely, typically constants ω^2, τ^2 exist such that:

(a) $\left\{\frac{1-\rho}{\omega^2} L_{[t\tau^2/(1-\rho)^2]}\right\} t = 0, 1, \ldots \xrightarrow{D} \{\bar{\xi}_t\} t \ge 0$ in $D[0, \infty)$ when $\rho \uparrow 1$;

(b) the stationary distributions converge as well, $\frac{1-\rho}{\omega^2} L \xrightarrow{D} \bar{\xi}$ where $\bar{\xi}$ is the limiting stationary exponential random variable for RBM, $\mathbb{P}(\bar{\xi} > u) = e^{-2u}$;

(c) the moments in (b) converge as well, $\mathbb{E}L^k \approx \frac{\omega^{2k}}{(1-\rho)^k} c_k$ where $c_k = \mathbb{E}\bar{\xi}^k = k!/2^k$. In particular, $\mathbb{E}L \approx \frac{\omega^2}{2(1-\rho)}$, $\mathbb{E}L^2 \approx \frac{\omega^4}{2(1-\rho)^2}$.

Noting that the score $\nabla \log f_\rho(Y)$ has mean zero, it is suggested from the structure of $\{(L_t, S_t)\}$ to introduce a two-dimensional process $\left\{(\bar{\xi}_t, \eta_t)\right\}_{t \geq 0}$ with second component $\{\eta_t\}$ evolving as standard Brownian motion $\{\tilde{\eta}_t\}$ but restarted at zero whenever $\{\bar{\xi}_t\}$ hits zero. Since $\{L_t\}$ and $\{S_t\}$ are typically dependent, $\{\bar{\xi}_t\}$ and $\{\eta_t\}$ will be so as well. For the precise construction, let $\{\tilde{\xi}_t\}$, $\{\tilde{\eta}_t\}$ denote Brownian motions which are defined on the same probability space, have unit variance, drift -1 for $\{\tilde{\xi}_t\}$ and zero for $\{\tilde{\eta}_t\}$, and covariance structure given by $\mathbf{Var}(\tilde{\eta}_t) = t(\gamma^2 + \delta^2)$, $\mathbf{Cov}(\bar{\xi}_t, \tilde{\eta}_t) = t\gamma$ where γ, δ are suitable constants whose values are derived from the variance–covariance structure of $\{(L_t, S_t)\}$ as explained in detail in [3]. Then we may represent RBM by means of the formula

$$\bar{\xi}_t = \tilde{\xi}_t - \min_{0 \leq u \leq t} \tilde{\xi}_u = \tilde{\xi}_t - \tilde{\xi}_{\nu(t)} \tag{2.6}$$

where $\nu(t) = \max\{u \in [0, t] : \tilde{\xi}_u \leq \tilde{\xi}_v, 0 \leq v \leq u\}$ is the time of the last minimum before t, and define $\eta_t = \tilde{\eta}_t - \tilde{\eta}_{\nu(t)}$.

Theorem 2.1 *Under appropriate regularity conditions, there exist constants $\omega^2, \tau^2, \gamma, \delta$ such that*

$$\left\{ \left(\frac{1-\rho}{\omega^2} L_{t\tau^2/(1-\rho)^2}, (1-\rho) S_{t\tau^2/(1-\rho)^2}\right) \right\} t = 0, 1, \ldots \xrightarrow{\mathcal{D}} \left\{(\bar{\xi}_t, \eta_t)\right\} t \geq 0$$

in $D[0, \infty)$. Furthermore,

$$\left(\frac{1-\rho}{\omega^2} L, (1-\rho) S\right) \xrightarrow{\mathcal{D}} (\bar{\xi}, \eta) \tag{2.7}$$

$$\mathbb{E}[L^k S^m] \approx \frac{\omega^{2k}}{(1-\rho)^{k+m}} \mathbb{E}[\bar{\xi}^k \eta^m]. \tag{2.8}$$

In order for this result to be useful (and in particular to reduce (2.8) to the explicit expressions (2.10), (2.11) below), we must determine the steady–state distribution of $\left\{(\bar{\xi}_t, \eta_t)\right\}$. By the *inverse Gaussian distribution function* $G(t; x)$ *with parameter x* we understand the distribution of the time of first passage of $\{\bar{\xi}\}$ from 0 to $-x$ (cf. e.g. [1, p. 263]).

Theorem 2.2 *The steady-state distribution of $\left\{(\bar{\xi}_t, \eta_t)\right\}$ is that of $(m, \tilde{\eta}_\omega)$ where $m = \max_{0 \leq t < \infty} \tilde{\xi}_t$ and $\omega = \inf\{t \geq 0 : \tilde{\xi}_t = m\}$. Here the marginal distribution of m is exponential with rate 2, the conditional distribution of ω given $m = x$ is inverse Gaussian with parameter x, and the conditional distribution of $\tilde{\eta}_\omega$ given $m = x, \omega = t$ is normal with mean $\gamma x + \gamma t$ and variance $\delta^2 t$. In particular,*

$$\mathbb{E}\eta = 2\gamma \mathbb{E}\bar{\xi} = \gamma, \quad \mathbb{E}\eta^2 = 4\gamma^2 \mathbb{E}\bar{\xi}^2 + (\delta^2 + \gamma^2)\mathbb{E}\bar{\xi} = \frac{5}{2}\gamma^2 + \frac{1}{2}\delta^2, \tag{2.9}$$

$$\mathbb{E}(\bar{\xi}^k \eta) = 2\gamma \mathbb{E}\bar{\xi}^{k+1} = 2\gamma c_{k+1}, \tag{2.10}$$

$$\mathbb{E}(\bar{\xi}^k \eta^2) = 4\gamma^2 \mathbb{E}\bar{\xi}^{k+2} + (\delta^2 + \gamma^2)\mathbb{E}\bar{\xi}^{k+1}$$
$$= 4\gamma^2 c_{k+2} + (\delta^2 + \gamma^2) c_{k+1}. \tag{2.11}$$

Turning back to (2.1) and its extension to higher order moments, we finally obtain the following main result stating how to compute approximative values for the sensitivities in heavy traffic:

Corollary 2.1 *Under conditions similar as for Theorem 2.1, the sensitivities* $\nabla \ell(\rho) = \nabla(\mathbb{E}_\rho L), \nabla(\mathbb{E}_\rho L^2), \ldots$ *satisfy*

$$\nabla(\mathbb{E}_\rho L) \approx \frac{\gamma \omega^2}{2(1-\rho)^2}, \quad \nabla(\mathbb{E}_\rho L^2) \approx \frac{\gamma \omega^4}{(1-\rho)^3}, \quad \ldots \tag{2.12}$$

$$\nabla(\mathbb{E}_\rho L^k) \approx \frac{\gamma \omega^{2k}(2c_{k+1} - c_k)}{(1-\rho)^{k+1}} = \frac{\gamma \omega^{2k} k! k}{2^k (1-\rho)^{k+1}} \tag{2.13}$$

3   The performance of the SF estimators

Turning to simulation, let N be the number of cycles simulated, $C^{(i)}$ the lenght of the ith cycle and let ti be shorthand for the tth customer in the ith cycle ($i = 1, \ldots, C^{(i)}$). Then the standard regenerative estimators motivated from (1.1) are

$$\hat{\ell} = \frac{\hat{r}}{\hat{c}}, \quad \text{where} \quad \hat{r} = \frac{1}{N} \sum_{i=1}^{N} \sum_{t=1}^{C^{(i)}} L_{ti}, \quad \hat{c} = \frac{1}{N} \sum_{i=1}^{N} C^{(i)}. \tag{3.1}$$

In the setting of sensitivity analysis, the expressions (2.2) and (2.3) suggest in a similar way two alternative simulation estimators for $\nabla r(\rho)$, namely

$$\nabla \hat{r}^{(\text{cr})}(\rho) = \frac{1}{N} \sum_{i=1}^{N} S_{C^{(i)}i} \sum_{t=1}^{C^{(i)}} L_{ti}, \tag{3.2}$$

$$\nabla \hat{r}^{(\text{eff})}(\rho) = \frac{1}{N} \sum_{i=1}^{N} \sum_{t=1}^{C^{(i)}} L_{ti} S_{ti}. \tag{3.3}$$

We refer to (3.2) and (3.3) as the *crude* and the *efficient* estimator, respectively. Similar notation is used for the estimators for $\nabla \ell(\rho)$ which we obtain by inserting (3.2), resp. (3.3) and the analogue estimators for $\nabla c(\rho)$ in (2.1).

Empirical evidence has suggested that $\nabla \hat{r}^{(\text{eff})}(\rho)$ and $\nabla \hat{\ell}^{(\text{eff})}(\rho)$ are the more efficient estimators. However, no theoretical support for this observation has been given (in fact, just the mathematical discussion of the relevant likelihood identities in the literature is not always rigorous). For a simple heuristic argument, note that (2.3) follows from (2.2) by evaluating the t–th term $S_C L_t$ by conditioning upon the history up to time t. Since conditioning reduces variance, we thus have $\text{Var}(L_t S_t) < \text{Var}(S_C L_t)$. However, somewhat surprisingly a counterexample was found in [3], showing that this inequality does not generalize to the whole sums in (2.2), (2.3): the covariances may mess up things.

To advocate the use of the efficient rather than the crude estimators, we thus need to invoke different types of argument, and our main vehicle will here be heavy traffic approximations (a different course is taken in [5]). The idea behind is the same as in [2], [9], to show that even if the evaluation of quantities like $\text{Var}\nabla \hat{\ell}^{(\text{cr})}(\rho)$, $\text{Var}\nabla \hat{\ell}^{(\text{eff})}(\rho)$ does not seem tractable (except maybe for some very simple special cases like M/M/1), then heavy-traffic approximations can be obtained. In the present setting, they permit in particular to conclude that the inequality

$$\text{Var}\nabla\hat{\ell}^{(\text{cr})}(\rho) > \text{Var}\nabla\hat{\ell}^{(\text{eff})}(\rho) \tag{3.4}$$

holds at least when ρ is sufficiently close to one and also to assess the value of the variance reduction

$$\frac{\text{Var}\nabla\hat{\ell}^{(\text{cr})}(\rho)}{\text{Var}\nabla\hat{\ell}^{(\text{eff})}(\rho)}. \tag{3.5}$$

Before stating our results it is, however, instructive to summarise the results for $\ell(\rho) = \mathbb{E}_\rho L$ that we are generalising. Assume that we want to estimate $\ell(\rho)$ by regenerative simulation using N cycles, cf. (3.1). Then the variance on the estimate is asymptotically of the form r^2/N for some constant r^2. Further, by standard heavy traffic theory (Section 2), $\mathbb{E}_\rho L = O((1-\rho)^{-2})$ when $\rho \uparrow 1$, and also the relation

$$r^2 \approx \frac{1}{\mathbb{E}_\rho C}\frac{v^2}{(1-\rho)^4} \tag{3.6}$$

has been part of the folklore for some time and is shown rigorously in [2] (cf. also the discussion in [9] and references there). Here the constant v^2 can be identified in terms of covariance constants for RBM and the numerical value is $1/2$. Note that typically $\mathbb{E}_\rho C = O((1-\rho)^{-1})$ so that (3.6) is $O((1-\rho)^{-3})$.

In sensitivity analysis, we are interested instead in estimating $\nabla\ell(\rho)$ which according to Section 2 is $O((1-\rho)^{-2})$. With N being the number of cycles simulated, we have again that

$$\text{Var}\nabla\hat{\ell}^{(\text{cr})}(\rho) \approx \frac{\sigma^2(\text{cr})}{N}, \quad \text{Var}\nabla\hat{\ell}^{(\text{eff})}(\rho) \approx \frac{\sigma^2(\text{eff})}{N}$$

for suitable constants, and our main result is the following:

Theorem 3.1 *There exist constants $s^2(\text{cr})$, $s^2(\text{cr})$ such that*

$$\sigma^2(\text{cr}) \approx \frac{1}{\mathbb{E}_\rho C}\frac{s^2(\text{cr})}{(1-\rho)^6}, \quad \sigma^2(\text{eff}) \approx \frac{1}{\mathbb{E}_\rho C}\frac{s^2(\text{eff})}{(1-\rho)^6}. \tag{3.7}$$

Unfortunately, this result is slightly more complicated here than (3.6) since $s^2(\text{cr})$, $s^2(\text{eff})$ depend on the constants γ^2, δ^2 associated with the correlation structure of the process (cf. Section 2). Numerical tables presented in [3] show, however, that typical values of the variance reduction $s^2(\text{cr})/s^2(\text{eff})$ are of the order of magnitude 2. Thus, the theorem provides a solution to our problem of providing quantitative measures for the difference between the crude and the efficient method.

Some further practical implications are briefly discussed in [3] and are closely related to [2], [9].

4   The performance of LR estimators: a M/M/1 study

In the study of the sensitivities, we assumed for simplicity that the reference parameter ρ_0 was taken as the governing parameter ρ of a single queueing system under study. We now turn to the case $\rho \neq \rho_0$, considering for simplicity only the steady state mean $\ell(\rho) = \mathbb{E}_\rho L$ and associated quantities like $r(\rho)$, and not the sensitivities.

Typically, absolute continuity is present in the sense that the support of the density $f_\rho(y)$ does not depend on ρ. Thus, the likelihood ratio for $L_1, \ldots, L_t$ is

$$W_t = W_t(\rho|\rho_0) = \prod_{i=1}^{t} \frac{f_\rho(Y_i)}{f_{\rho_0}(Y_i)} \tag{4.1}$$

so that

$$r(\rho) = \mathbb{E}_{\rho_0} W_C R = \mathbb{E}_{\rho_0} W_C \sum_{t=1}^{C} L_t, \quad \ell(\rho) = \frac{\mathbb{E}_{\rho_0} W_C R}{\mathbb{E}_{\rho_0} W_C C} = \frac{\mathbb{E}_{\rho_0} W_C \sum_{t=1}^{C} L_t}{\mathbb{E}_{\rho_0} W_C \sum_{t=1}^{C} 1} \tag{4.2}$$

and an alternative to generate

$$\hat{r}(v) = \frac{1}{N} \sum_{i=1}^{N} W_{Ci} \sum_{t=1}^{C_i} L_{ti}$$

by simulation using the parameter ρ is to simulate

$$\hat{r}(v|v_0) = \frac{1}{N} \sum_{i=1}^{N} W_{Ci} \sum_{t=1}^{C_i} L_{ti}, \tag{4.3}$$

using the parameter ρ_0. In [4] it is shown that one can replace this estimator with one of the form

$$\hat{r}(v|v_0) = \frac{1}{N} \sum_{i=1}^{N} \sum_{t=1}^{C_i} W_{ti} L_{ti}, \tag{4.4}$$

that is, 'W_C can be moved under the sum sign as W_t'. Empirical evidence (simulations) and parallels to Section 3 indicate that (4.4) is typically the more efficient version, and the following discussion is based on estimators of this type alone. Thus, with (4.4) at hand, our estimator for $\ell(\rho)$ is

$$\hat{\ell}(\rho|\rho_0) = \frac{\sum_{i=1}^{N} \sum_{t=1}^{C_i} W_{ti} L_{ti}}{\sum_{i=1}^{N} \sum_{t=1}^{C_i} W_{ti}} \tag{4.5}$$

It is crucial to realize that estimators of the type (3.1) allow to estimate $r(\rho)$ for a *fixed* ρ only, while the counterpart (4.5) allows to estimate the same parameter for *different* values of ρ by simulating a *single* scenario under ρ_0. Similar remarks apply to the estimators $\hat{\ell}(\rho|\rho_0)$ compared to the standard regenerative estimator of $\ell(\rho)$. In this way one obtains a solution of what we call (e.g. [7]) the *"what if"* problem, which can be formulated as follows: What will be the values of $\ell(\rho)$ if we perturb the parameter ρ by $\Delta\rho = \rho_0 - \rho$?

We shall call the $\hat{\ell}(\rho|\rho_0)$ the *"what if"* estimators. It is worthwhile to note that they resemble importance sampling (IS) in the sense that both are based on a change of probability measure. They differ, however, in their main goals: the IS estimators are mainly introduced with the view of *variance reduction*, while the "what if" estimators are introduced with the view of *simultaneous estimation* of a number of performance measures while using *a single simulation run*. As we shall see below, in typical situations the "what

8

if" estimators still allow to obtain variance reduction (as compared to the crude regenerative method). This, of course, may be considered as a useful additional property of the "what if" estimators.

The aspect of simultaneous estimation appears particularly appealing for stochastic optimization purposes. For example, a standard minimization problem is

$$\text{Minimize } \{G(\rho) = \mathbb{E}_\rho L + \varphi(\rho)\}, \quad \rho \in \Theta \subset (0,1),$$

where $\varphi(\rho)$ is a continuously differentiable function of ρ, cf. [7]. For an illustration of the aspect of simultaneous estimation, see Fig. 2, where we are considering the expected sojourn time of the M/M/1 queue with arrival rate $\beta = 1$ and service rate $1/\rho$. We simulated a single sample path corresponding to 10.000 customers with reference parameter $\rho_0 = 0.9$ and produced the figure by computing $\ell(\rho)$, $\hat{\ell}(\rho|\rho_0)$ and the corresponding 95% confidence interval.

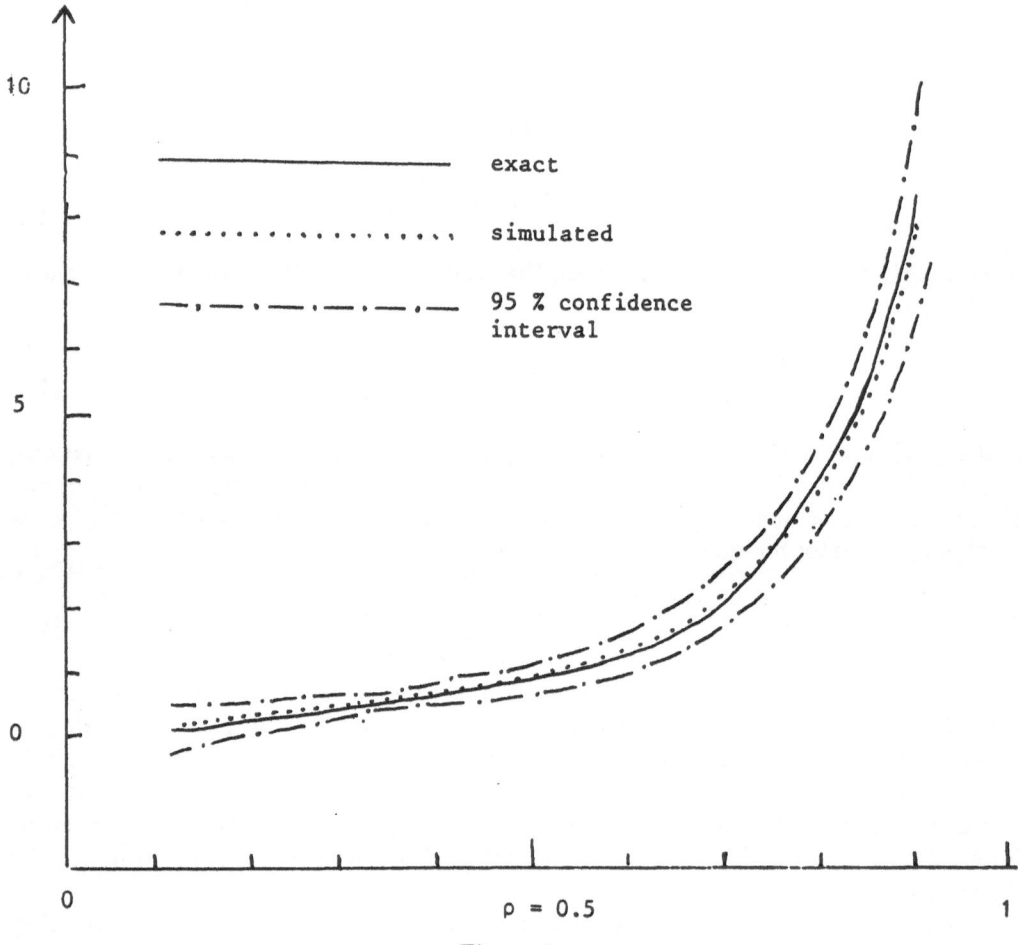

Figure 2:

Very little seems to be known about the behavior of "what if" estimators, even from the empirical point of view. We shall here present a study (extracted from [4]) for the special case where L_t is the waiting time of customer t in a M/M/1 queue with traffic intensity ρ. The crucial feature of the M/M/1 model is that it allows to compute explicitly its basic characteristics like $\sigma_L^2(\rho|\rho_0)$, the variance constant in the asymptotic $\mathcal{N}(\ell(\rho), N^{-1}\sigma_L^2(\rho|\rho_0))$ distribution of $\hat{\ell}(\rho|\rho_0)$; note that the heavy traffic point of view may be less appropriate here since ρ may differ substantially from ρ_0 so that both need not be close to 1.

Our first finding was that $\sigma_L^2(\rho|\rho_0)$ is finite for

$$\rho \leq \rho_c = \rho_c(\rho_0) = \frac{2\rho_0}{1 + \sqrt{\rho_0}} \tag{4.6}$$

and infinite for $\rho > \rho_c$. Take for example $\rho_0 = 0.1$ and 0.9. Then ρ_c=0.152 and 0.924, respectively (for further illustration, see Fig. 9 below). Thus, if the reference parameter ρ_0 is only slighty larger ρ, the variance is infinite, and confidence intervals loose their meaning.

In the following, three values of the traffic intensity, namely $\rho = 0.3$, 0.6, and 0.9 were specified. We used them either for ρ or ρ_0 to illustrate light, moderate and heavy traffic, respectively.

Define

$$\epsilon(\rho|\rho_0) = \frac{\sigma_L^2(\rho|\rho_0)t(\rho_0)}{\sigma_L^2(\rho|\rho)t(\rho)} \tag{4.7}$$

as the efficiency of the "what if" estimator $\hat{\ell}(v|v_0)$ relative to the standard (crude Monte Carlo) one $\hat{\ell}(v) = \hat{\ell}(v|v_0)$. Here $t(\rho_0)$ and $t(\rho)$ present the CPU time required to compute the "what if" and the standard estimators respectively. We assume that $t(\rho) = \mathbb{E}_v C$ and similarly $t(\rho_0) = \mathbb{E}_{v_0} C$. By doing so we shall measure the efficiency in terms of variance/customer rather than variance/cycle.

Figs. 3, 4, 5 present the relative efficiency $\epsilon(\rho|\rho_0)$ as a function of ρ for the selected values 0.3, 0.6, 0.9. It is seen from the results of the figures that:

- There exists a rather broad ρ-interval of the form (ρ_n, ρ_0) where some moderate variance reduction is achieved, that is $\epsilon(\rho|\rho_0) < 1$.

- The efficiency $\epsilon(\rho|\rho_0)$ decreases moderately, $(\epsilon(\rho|\rho_0) \geq 1)$ as we move from ρ_0 down to zero, and it decreases $(\epsilon(\rho|\rho_0) \gg 1)$ as we move above ρ_0

In Figs. 6, 7, 8 the question has been turned around, in the sense that they display how different values of ρ_0 perform for the selected values 0.3, 0.6, 0.9 of ρ. Put together with the above discussion, the overall picture emerges that our "what if" estimators perform well in the range $\rho_n(\rho_0) = \rho_n < \rho \leq \rho_0$ (the *variance reduction region*). Moving away from this region variances blow up rather rapidly, especially above ρ_0. The shape of the variance reduction region and the optimal value function $\rho_0^*(\rho)$ are depicted in Fig. 9. It follows from Fig. 9 that we have good performance ($\epsilon(\rho|\rho_0) < 1$) in a reasonably large neighborhood of $\rho_0^*(\rho)$. This clearly indicates that it is desirable to choose the reference parameter ρ_0 moderately larger than the underlying parameter ρ. We consider this one of our main findings, and would like to add that although the M/M/1 set-up is quite specialized, nevertheless we believe that our results shed some light and provide a basic insight on more complicated queueing models, at least from a qualitative point of view.

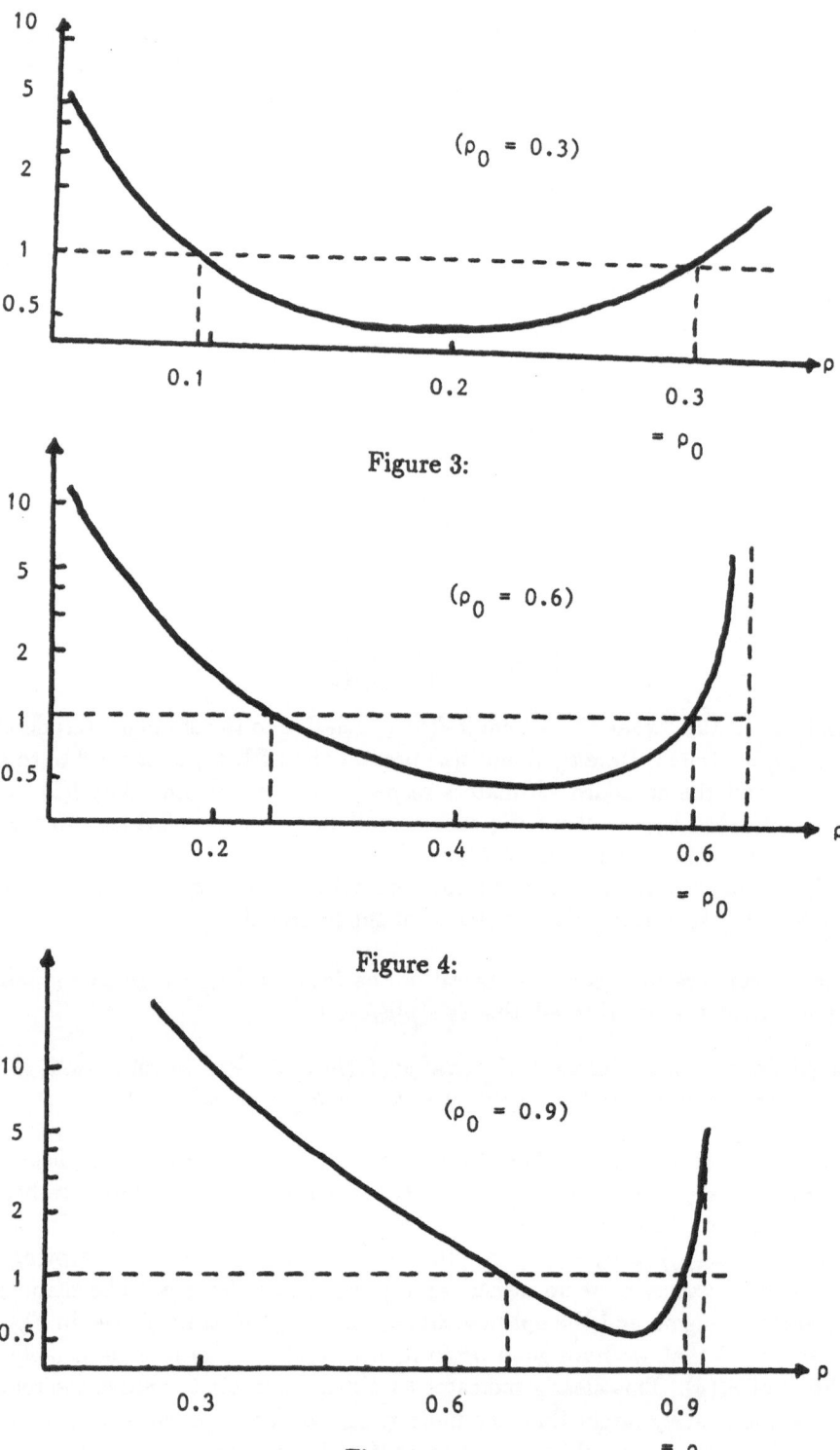

Figure 3:

Figure 4:

Figure 5:

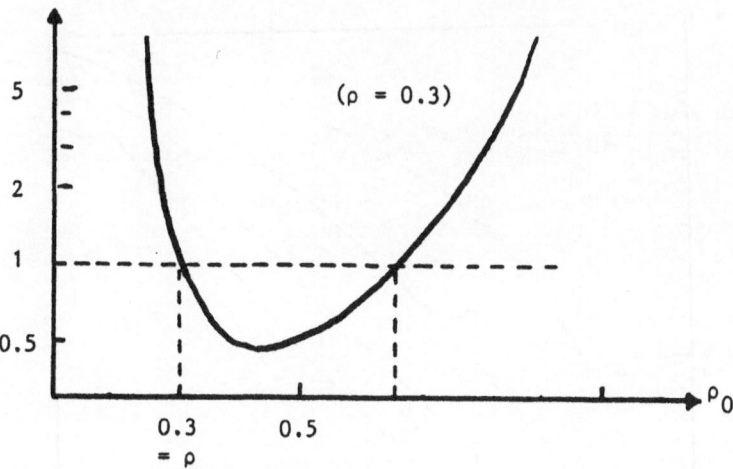

Figure 6:

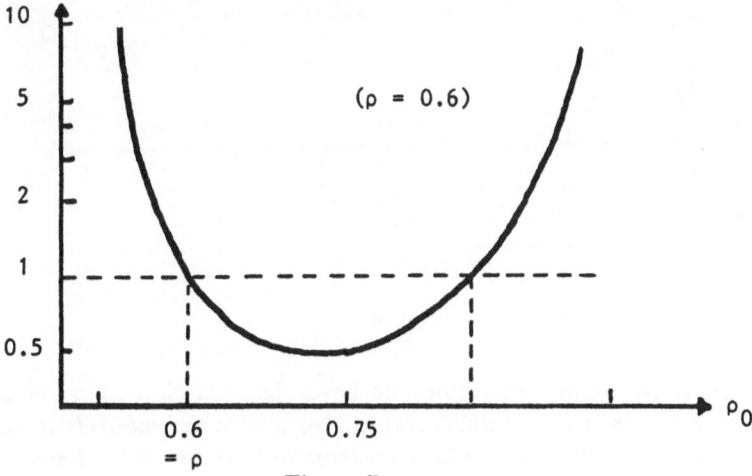

Figure 7:

Figure 8:

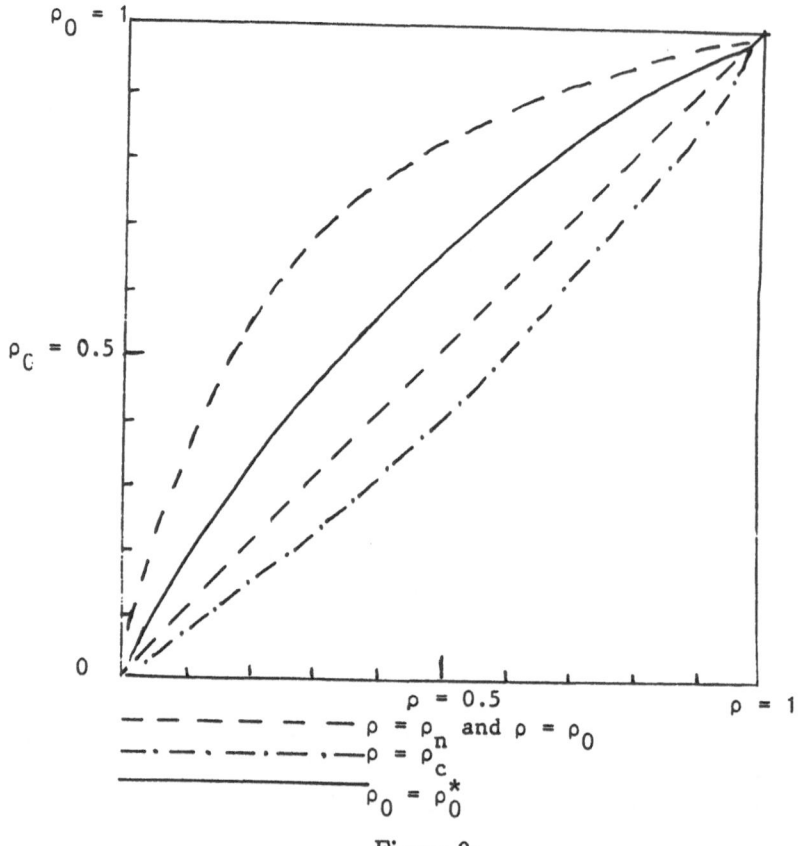

Figure 9:

In fact, from extensive numerical studies we have observed that many of our M/M/1 findings, and in particular the one which states: *"we are on the safe side if we choose the reference parameter v_0 corresponding to a moderately larger traffic intensity"*, are valid also for tandem queues and simple networks.

It seems reasonable to ask whether there is a simple intuitive explanation that for a given ρ one can obtain variance reduction by choosing ρ_0 moderately larger than ρ. We suggest to answer this in terms of importance sampling: when estimating $\ell(\rho)$, the main contribution is typically from cycles which are somewhat larger than average (for example in heavy traffic, essentially only cycles of the order of magnitude $(1-\rho)^{-2}$ matter whereas the cycles themselves are of the order of magnitude $O(1)$ in distribution, (see the discussion of [2]); and choosing a reference parameter $\rho_0 > \rho$ puts more weight on large cycles.

5 Extensions

The set–up and many of the results of the present paper allows for many generalizations, in particular the following (see [3], [4]):

- The parameter governing the queueing process $\{L_t\}$ needs not coincide with the traffic intensity and may be multi–dimensional;

- The input sequence $\{Y_i\}$ needs not be i.i.d. and may be vector–valued;

- We may consider the higher–order sensitivities $\nabla^p \ell(\rho)$, e.g. the Hessian $\nabla^2 \ell(\rho)$;

- Using a suitable control variate procedure, one can combine the crude and the efficient estimators to a further one $\nabla \hat{\ell}^{(\text{ctr})}(\rho)$ which is even more efficient.

References

[1] S. Asmussen (1987) *Applied Probability and Queues*, John Wiley & Sons.

[2] S. Asmussen (1991) Queueing simulation in heavy traffic. *Math. Opns. Res.*

[3] S. Asmussen & R.Y. Rubinstein (1992) The efficiency and heavy traffic properties of the score function method for sensitivity analysis of queueing models. *Adv. Appl. Probab.*

[4] S. Asmussen & R.Y. Rubinstein (1991) The performance of likelihood ratio estimators using the score function. *Stoch. Models.* (under revision).

[5] P. Glasserman (1990) Stochastic monotonicity, total positivity, and conditional Monte Carlo for likelihood ratios. Manuscript, AT& T Bell Laboratories, Holmdel, N.J.

[6] R.Y. Rubinstein (1989) Sensitivity analysis of computer simulation models via the efficient score. *Opns.Res.* **37**, 72-81.

[7] R.Y. Rubinstein (1989b) Monte Carlo Methods for performance evaluation, sensitivity analysis and optimization of stochastic systems, *Encyclopedie of Computer Science and Technology* (Kent ed.) (to appear).

[8] R.Y. Rubinstein & A. Shapiro (1991) *Discrete Event Systems: Sensitivity Analysis and Stochastic Optimization via the Score Function Method*, to be published by John Wiley & Sons.

[9] W. Whitt (1989) Planning queueing simulations. *Managmt. Sci.* **35**, 1341-1366.

EXPERIMENTAL RESULTS FOR GRADIENT ESTIMATION
AND OPTIMIZATION OF A MARKOV CHAIN IN STEADY-STATE

Pierre L'Ecuyer
Département d'I.R.O., Université de Montréal,
C.P. 6128, Montréal, H3C 3J7, Canada

Nataly Giroux
Département d'informatique, Université Laval,
Ste-Foy, Québec, G1K 7P4, Canada

Peter W. Glynn
Operations Research Department, Stanford University,
Stanford, CA94305, U.S.A.

ABSTRACT. Infinitesimal perturbation analysis (IPA) and the likelihood ratio (LR) method have drawn lots of attention recently, as ways of estimating the gradient of a performance measure with respect to continuous parameters in dynamic stochastic systems. In this paper, we experiment with the use of these estimators in stochastic approximation algorithms, to perform so-called "single-run optimizations" of steady-state systems, as suggested in [23]. We also compare them to finite-difference estimators, with and without common random numbers. In most cases, the simulation length must be increased from iteration to iteration, otherwise the algorithm converges to the wrong value. We have performed extensive numerical experiments with a simple M/M/1 queue. We state convergence results, but do not give the proofs. The proofs are given in [14].

1. THE MODEL AND THE STOCHASTIC APPROXIMATION APPROACH

We consider a Markov chain $\{X_i(\theta, s, \omega), i = 0, 1, \ldots\}$ with Borel state space S, defined over a probability space $(\Omega, \Sigma, P_{\theta,s})$. Its associated probability measure $P_{\theta,s}$ depends on the parameter vector θ, where $\theta \in \bar{\Theta} \subset \mathbb{R}^d$, and on the (deterministic) initial state $X_0(\theta, s, \omega) = s \in S$. A cost $f(\theta, x)$ is incurred whenever we visit state x (except for the initial state X_0), for measurable $f : \bar{\Theta} \times S \to \mathbb{R}$. Let

$$h_t(\theta, s, \omega) = \frac{1}{t} \sum_{i=1}^{t} f(\theta, X_i(\theta, s, \omega)) \tag{1}$$

be the average cost for the first t steps and

$$\alpha_t(\theta, s) = \int_{\Omega} h_t(\theta, s, \omega) \mathrm{d}P_{\theta,s}(\omega), \tag{2}$$

its expectation. Let $\bar{S}$ be a subset of S which can be viewed as the set of admissible initial states. For example, $\bar{S}$ can consist of a single state s_0 if all simulation subruns (below) are started from that state. In most other cases, $\bar{S}$ will be a compact subset of S. We assume that

$$\lim_{t \to \infty} \sup_{\theta \in \Theta, s \in S} |\alpha_t(\theta, s) - \alpha(\theta)| = 0, \tag{3}$$

where $\alpha(\theta)$ is the steady-state average cost for running the system at parameter level θ, and that

$$\lim_{l \to \infty} \sup_{\theta \in \Theta, s \in S} |\nabla_\theta \alpha_l(\theta, s) - \nabla_\theta \alpha(\theta)| = 0. \tag{4}$$

We assume that each $\alpha_l(\cdot, s)$ and α are differentiable. Suppose one is interested in minimizing $\alpha(\theta)$ over Θ, a compact and convex subset of $\bar{\Theta}$ such that each point of Θ has a neighborhood inside $\bar{\Theta}$. Let $\theta^* \in \Theta$ be the optimum (assumed to be unique). We consider a stochastic approximation (SA) algorithm of the form

$$\theta_{n+1} := \pi_\Theta(\theta_n - \gamma_n Y_n). \tag{5}$$

for $n \geq 1$, where θ_n is the parameter value at the beginning of iteration n ($\theta_1 \in \Theta$ is random with known distribution), Y_n is an estimate of the gradient $\nabla \alpha(\theta_n)$ obtained at iteration n, $\{\gamma_n, n \geq 1\}$ is a (deterministic) positive sequence decreasing to 0 and such that $\sum_{n=1}^\infty \gamma_n = \infty$, and π_Θ denotes the projection on the set Θ. In what follows, except when stated otherwise, we will assume that $\gamma_n = \gamma_0 n^{-1}$ for some constant $\gamma_0 > 0$.

Each Y_n is obtained by simulating the system for one or more subrun(s) of finite duration. Each simulation subrun corresponds essentially to one copy of the Markov chain described above, with initial state $s \in \bar{S}$. Specific ways of obtaining Y_n can be based, for example, on IPA, LR, or finite differences. We recall them in the next section. In general, since Y_n must be obtained in finite time, it is a biased estimator of $\nabla \alpha(\theta)$.

Let $s_n \in \bar{S}$ denote the *state* of the system at the beginning of iteration n. In the simulation program, this corresponds to the description of all the objects in the system, event list, etc.. We assume that when θ_n and s_n are fixed, the distribution of (Y_n, s_{n+1}) is completely specified and independent of the past iterations (but may depend on n). Here, Y_0 is a dummy value. Denote by $E_n(\cdot)$ the conditional expectation $E(\cdot \mid \theta_n, s_n)$, that is the expectation conditional on what is known at the beginning of iteration n. Assume that Y_n is integrable for all $n \geq 1$ and let:

$$Y_n = \nabla \alpha(\theta_n) + \beta_n + \epsilon_n \tag{6}$$

where $\beta_n = E_n[Y_n] - \nabla \alpha(\theta_n)$ represents the (conditional) bias on Y_n given (θ_n, s_n), while ϵ_n is a random noise, with $E_n(\epsilon_n) = 0$. The next proposition gives simplified sufficient conditions for the convergence of (5) to an optimum. More general conditions are given, e.g., in [10, 11, 14, 16, 18].

PROPOSITION 1. *Let $d = 1$. Θ is now a closed interval of $\mathbb{R}$. Assume that $\alpha(\theta)$ is strictly unimodal over $\bar{\Theta}$. If $\lim_{n \to \infty} \beta_n = 0$ and $\sum_{n=1}^\infty E_1(\epsilon_n^2) n^{-2} < \infty$ with probability one, then $\lim_{n \to \infty} \theta_n = \theta^*$ with probability one.* ∎

2. Gradient estimators

2.1. Finite differences (FD)

Here, we consider *central* FD. For other variants, like *forward* FD, see [6, 10, 18]. For simplicity, let $d = 1$. Take a deterministic positive sequence $\{c_n, n \geq 1\}$ that converges to 0. At iteration n, simulate from some initial state $s_n^- \in \bar{S}$ at parameter value $\theta_n^- = \pi_\Theta(\theta_n - c_n)$ for t_n transitions. Simulate also (independently) from state $s_n^+ \in \bar{S}$ at parameter value $\theta_n^+ = \pi_\Theta(\theta_n + c_n)$ for t_n transitions. Let ω_n^- and ω_n^+ denote the respective sample points. The FD gradient estimator is

$$Y_n = \frac{h_{t_n}(\theta_n^+, s_n^+, \omega_n^+) - h_{t_n}(\theta_n^-, s_n^-, \omega_n^-)}{\theta_n^+ - \theta_n^-}. \tag{7}$$

For $d > 1$, just repeat the procedure for each component of θ_n.

Here, there are different sources of bias: there is bias due to the fact that we just simulate over a finite horizon, bias due the finite differences, and bias due to the possibly different initial states s_n^- and s_n^+. To get $\beta_n \to 0$, one typically needs to take $t_n \to \infty$. In the light of [6, 10], where related problems are discussed, reasonable choices for the sequences might be for instance $t_n = t_a + t_b n$ and $c_n = c_0 n^{-1/6}$ for appropriate constants t_a, t_b, and c_0. One simple way to choose the initial states of the subruns is as follows. Start the first subrun of iteration n from state $s_n \in \tilde{S}$, then take the terminal state of any given subrun as the initial state of the next one. (Project on $\tilde{S}$ whenever necessary.) For s_{n+1}, take the terminal state of the last subrun of iteration n. Another way is to take the same initial state for each subrun: $s_n^- = s_n^+ = s_n$. One can also take (reset) $s_n = s_0$ for all n, for a fixed state s_0. In any case, this method is usually plagued by a huge variance on Y_n, which makes it converge very slowly, at least when the subruns are performed with "independent" random numbers.

2.2. Finite differences with common random numbers (FDC)

One way to reduce the variance in FD is to use common random numbers across the subruns at each iteration, start all the subruns from the same state: $s_n^- = s_n^+ = s_n$, and synchronize. More specifically, one views ω as representing a sequence of $U(0,1)$ variates, so that all the dependency on (θ, s) appears in $h_t(\theta, s, \cdot)$. Take $\omega_n^+ = \omega_n^- = \omega_n$. Since the subruns are aimed at comparing very similar systems, $h_{t_n}(\theta_n^+, s_n, \omega_n)$ and $h_{t_n}(\theta_n^-, s_n, \omega_n)$ should be highly correlated, especially when c_n is small, so that considerable variance reductions might be obtained. Conditions that *guarantee* variance reductions are given in [3, 18]. In practice, this method is not always easy to implement. See the discussion in [14]. Reasonable choices for the sequences are $t_n = t_a + t_b n$ and $c_n = c_0 n^{-1/5}$ for appropriate constants t_a, t_b, and c_0. This is somewhat justified by the results of [6], where the related finite-horizon gradient estimation problem is analyzed.

2.3. A likelihood ratio (LR) approach

With the LR approach [1, 5, 7, 12, 17, 19, 20], to differentiate the expectation (2) with respect to θ, we first take a probability measure G independent of θ that dominates the $P_{\theta,s}$'s for $\theta \in \bar{\Theta}$, $s \in \tilde{S}$, and rewrite:

$$\alpha_t(\theta, s) = \int_\Omega h_t(\theta, s, \omega) L(G, \theta, s, \omega) dG(\omega), \tag{8}$$

where $L(G, \theta, s, \omega) = (dP_{\theta,s}/dG)(\omega)$ is the *Radon-Nikodym* derivative of $P_{\theta,s}$ with respect to G. Under appropriate regularity conditions (see [12]), one can differentiate α_t by differentiating inside the integral:

$$\nabla \alpha_t(\theta, s) = \int_\Omega \psi_t(\theta, s, \omega) dG(\omega). \tag{9}$$

where

$$\psi_t(\theta, s, \omega) = L(G, \theta, s, \omega) \nabla_\theta h_t(\theta, s, \omega) + h_t(\theta, s, \omega) \nabla_\theta L(G, \theta, s, \omega). \tag{10}$$

When (9) holds, $\psi_t(\theta, s, \omega)$ can be used to estimate $\nabla \alpha_t(\theta, s)$.

Typically, ω can be viewed as the set of values taken by a finite sequence of independent random variables. For example, let $\omega = (\xi_1, \ldots, \xi_t)$, where for $1 \leq i \leq t$, ξ_i is the value taken by a continuous random variable (or vector) with density $g_{i,\theta}$. Given X_{i-1}, the value of ξ_i determines the next state X_i (i.e. X_i is a function of (X_{i-1}, ξ_i)). To estimate $\nabla \alpha_t(\theta_n, s)$, an easy choice for G is $P_{\theta_n, s}$. Then, the Radon-Nikodym derivative is the *likelihood ratio*

$$L(P_{\theta_n, s}, \theta, s, \omega) = \prod_{i=1}^t \frac{g_{i,\theta}(\xi_i)}{g_{i,\theta_n}(\xi_i)} \tag{11}$$

and its gradient is the *score function*:

$$S(\theta, s, \omega) = \sum_{i=1}^{t} \nabla_\theta \ln g_{i,\theta}(\xi_i). \tag{12}$$

A major problem is that the variance of $S(\theta, s, \omega)$ (and of the LR gradient estimator) typically increases linearly with t. In practice, there is a tradeoff between bias and variance: t_n must be increased with n, but not too fast.

When the system possesses a readily identifiable regenerative structure, α can be written as the quotient of two functions, and a LR approach can be used to obtain an estimator of the derivative of the quotient, for each component of θ. See [5, 7, 17] for more details. There is still a bias problem and t_n must still go to infinity, because that approach involves the expectation of a ratio, but the variance now decreases linearly instead of increasing with the simulation length. That approach could be practical if the regenerative cycles are not too long.

2.4. Infinitesimal Perturbation analysis (IPA)

Here, we define the sample space in such a way that $P_{\theta,s}$ is independent of θ. For instance, one can view ω as a sequence of independent $U(0,1)$ variates. Then, $L(P_{\theta_n, s_n}, \theta, s, \omega) = 1$ and (10) becomes:

$$\psi_t(\theta, s, \omega) = \nabla_\theta h_t(\theta, s, \omega). \tag{13}$$

This is the usual IPA gradient estimator for $\nabla \alpha_t(\theta, s)$ [8, 9, 21, 22].

3. A GI/G/1 QUEUE

Consider a GI/G/1 queue [2] with interarrival and service-time distributions A and B_θ respectively, both with finite expectations and variances. The latter depends on a parameter $\theta \in \tilde{\Theta} = [\ell_1, \ell_2] \subset \mathbb{R}$ and has a corresponding density function b_θ. We assume that for all $\theta \in \tilde{\Theta}$, the system is stable. Let $w(\theta)$ be the average sojourn time in the system per customer, in steady-state, at parameter level θ. The objective function is defined by

$$\alpha(\theta) = w(\theta) + C(\theta). \tag{14}$$

where $C : \tilde{\Theta} \mapsto \mathbb{R}$. We want to minimize $\alpha(\theta)$ (assumed strictly unimodal) over $\Theta = [a, b]$, where $\ell_1 < a < b < \ell_2$. For many distributions, $\alpha(\theta)$ and its minimizer θ^* can be computed analytically or numerically. But let us ignore this momentarily and try to solve the problem using SA, combined with different gradient estimation methods. The solutions of numerical examples can then be compared to the true optimal solutions for an empirical evaluation.

A GI/G/1 queue can be described in terms of a discrete-time Markov chain via Lindley's equation. Let W_i, ζ_i, and $X_i = W_i + \zeta_i$ be the *waiting* time, *service* time, and *system* time for the i-th customer, and ν_i be the time between arrivals of the $(i-1)$-th and i-th customer (for $i = 1$, it is the time until the first arrival). For our purposes, the system time X_i will be the state of the chain at step i. The state space is $S = [0, \infty)$ and $X_0 = s$ is the initial state. $X_0 = 0$ corresponds to an initially empty system. For $i \geq 0$, one has

$$X_i := (X_{i-1} - \nu_i)^+ + \zeta_i \tag{15}$$

where x^+ means $\max(x, 0)$. Since $C(\theta)$ is deterministic, we will estimate only the derivative of $w(\theta)$ and then add $C'(\theta)$ separately to Y_n. Therefore, here, the notation differs slightly from that of the previous sections: $f(\theta, X_i) = X_i$ and $h_t(\theta, s, \omega)$ represents the *average* system time for the t customers who leave during a given subrun of length t (customers). Let $\tilde{S} = [0, c]$ for some (perhaps large) constant c. Let $w_t(\theta, s) = E_{\theta,s}[h_t(\theta, s, \omega)]$ and $\alpha_t(\theta, s) = w_t(\theta, s) + C(\theta)$. We assume that α is continuously differentiable and strictly unimodal in $\tilde{\Theta}$. We also need the following assumptions.

ASSUMPTION 1.

(i) The set $\{\zeta \geq 0 \mid b_\theta(\zeta) > 0\}$, which is the support of b_θ, is independent of θ. Call it Δ.

(ii) Everywhere in $\bar{\Theta}$, $b_\theta(\zeta)$ is continuously differentiable with respect to θ, for each $\zeta \geq 0$, and continuous in ζ.

(iii) For each θ in $\bar{\Theta}$, b_θ has a finite Laplace transform in a neighborhood of zero.

(iv) For each θ_0 in $\bar{\Theta}$, $\lim_{\theta \to \theta_0} \sup_{\zeta \in \Delta} b_\theta(\zeta)/b_{\theta_0}(\zeta) = 1$.

(v) For each θ_0 in $\bar{\Theta}$, there exists $\Psi_{\theta_0} : (0,\infty) \to \mathbb{R}$ and a neighborhood Υ_{θ_0} of θ_0 such that $\sup_{\theta \in \Upsilon_{\theta_0}} |\partial b_\theta(\zeta)/\partial \theta|/b_{\theta_0}(\zeta) \leq \Psi_{\theta_0}(\zeta)$ for all ζ, and $\sup_{\theta_0 \in \Theta} E_{\theta_0}[\Psi_{\theta_0}^4(\zeta)] < \infty$. ∎

ASSUMPTION 2.

(i) Suppose that ζ_j is generated by inversion [3]: $\zeta_j = B_\theta^{-1}(U_j) \stackrel{\text{def}}{=} \inf\{\zeta \mid B_\theta(\zeta) \geq U_j\}$, where U_j is a $U(0,1)$ variate.

(ii) There is a distribution $\check{B}$ such that $\sup_{\theta \in \Theta} B_\theta^{-1}(u) \leq \check{B}^{-1}(u)$ for each u. The queue remains stable when the service times are generated according to $\check{B}$. Also, $\int_0^1 (\check{B}^{-1}(u))^4 du < \infty$ (finite second moment).

(iii) $B_\theta^{-1}(u)$ is differentiable in θ for each $u \in (0,1)$.

(iv) There exists a measurable function $\Gamma : (0,1) \mapsto \mathbb{R}$ such that $\sup_{\theta \in \Theta} |\partial B_\theta^{-1}(u)/\partial \theta| \leq \Gamma(u)$ for each $u \in (0,1)$ and such that $\int_0^1 (\Gamma(u))^4 du < \infty$. ∎

In [14], we prove that under Assumption 1, each $\alpha_i(\cdot, s)$ is continuously differentiable, and that under Assumption 2 and a mild additional condition, the uniform convergence conditions (3–4) hold. Note that some of these assumptions can be relaxed, but at the cost of getting more complicated.

In the context of this example, t_n represents the number of customers for each subrun of iteration n, except for the regenerative methods, where it represents the number of regenerative cycles per subrun. Regeneration points occur at the beginning of busy periods, i.e. when $X_i = s_0 = 0$. The following propositions are proven in [14].

PROPOSITION 2. *Consider SA with FD, with $t_n \to \infty$, $c_n \to 0$, $\sum_{n=1}^{\infty} t_n^{-1}(nc_n)^{-2} < \infty$, and $s_n^- = s_n^+$ for all n. Suppose Assumptions 1 and 2 and (3–4) hold. Then, $\theta_n \to \theta^*$ with probability one.*

For LR, one can view ω as representing the set of all interarrival and service times generated during a given subrun. One gets the score function (12) with $\xi_i = (\nu_i, \zeta_i)$ and, since only the service times depend on θ, one obtains:

$$Y_n = C'(\theta_n) + h_i(\theta, s, \omega) \sum_{i=1}^{i} \frac{\partial}{\partial \theta} \ln b_\theta(\zeta_i). \tag{16}$$

For the regenerative case, one has

$$Y_n = C'(\theta_n) + \left(\sum_{j=1}^{t_n} \tau_j \sum_{j=1}^{t_n} h_j S_j - \sum_{j=1}^{t_n} h_j \sum_{j=1}^{t_n} \tau_j S_j \right) \left(\sum_{j=1}^{t_n} \tau_j \right)^{-2} \tag{17}$$

where for $j = 1, \ldots, t_n$, τ_j is the number of departures during the j-th regenerative cycle, h_j is the total system time for those τ_j customers who left during that cycle, and $S_j = \sum_{i=1}^{\tau_j} \partial \ln b_\theta(\zeta_i)/\partial \theta$.

PROPOSITION 3. *Suppose that Assumption 1 and (3-4) hold, that $\sup_{\theta \in \Theta} E_\theta[\zeta^\delta + \Psi_\theta^\delta(\zeta)] < \infty$, and that one uses SA with LR as described above. With the truncated horizon approach, if $s_n \in \tilde{S}$ for all n, $t_n \to \infty$, and $\sum_{n=1}^\infty t_n n^{-2} < \infty$, then $\theta_n \to \theta^*$ with probability one. With the regenerative approach, if $t_n \to \infty$, then $\theta_n \to \theta^*$ with probability one.*

For fixed t, one can *decompose* the cost and estimate the gradient of the waiting time of each individual customer separately. The score function $S_t^{(i)}(\theta, s, \omega)$ associated to a given customer i need not be the sum of all t terms as above, but could include only the terms that correspond to the ζ_j's which can influence that customer's system time. It is

$$S_t^{(i)}(\theta, s, \omega) = \sum_{j=1}^i \frac{\partial}{\partial \theta} \ln b_\theta(\zeta_j)$$

and the estimator of $\nabla \alpha_t(\theta, s)$ becomes

$$C'(\theta) + \frac{1}{t} \sum_{i=1}^t X_i \sum_{j=1}^i \frac{\partial}{\partial \theta} \ln b_\theta(\zeta_j). \tag{18}$$

This LR estimator has approximately half the number of terms as the previous one. Another way of reducing variance is to use the estimator $C'(\theta) + (1/t) \sum_{i=1}^t (X_i - w(\theta)) S_t^{(i)}(\theta, s, \omega)$, in which $w(\theta)$, when unknown, can be replaced by an independent estimate. See [13] for further details.

For IPA [12, 21, 23], the idea is to differentiate (15) for a fixed set of underlying uniform variates. An infinitesimal perturbation on ζ_j affects the system time of customer j and of all the customers (if any) following him in the same busy period. Therefore,

$$\nabla_\theta h_t(\theta, s, \omega) = \frac{1}{t} \sum_{i=1}^t \sum_{j \in \Xi_i} \frac{\partial B_\theta^{-1}(U_j)}{\partial \theta} \tag{19}$$

where Ξ_i is the set containing customer i and all the customers that precede him in the same busy period (if any). We call the inside sum in (19) the *IPA accumulator*. When the state is not reset to s_0 between iterations, we can consider two variants of this: one in which Ξ_i can contain only customers who left during the current iteration (the IPA accumulator is reset to zero between iterations) and one in which it can contain customers from the previous iterations (which have indices $j \leq 0$ in (19)). In the latter case, the value a_n of the IPA accumulator at the beginning of iteration n must be stored and could be viewed as part of s_n. For IPA with a regenerative approach, the estimator is defined as in (19), but with t denoting the number of customers that leave during the t_n regenerative cycles. In that case, t is the value taken by a random variable T_n.

PROPOSITION 4. *Suppose that (3-4) hold and that one uses SA with IPA, under Assumptions 1 and 2, with $s_1 = s_0$ and with $t_n \to \infty$. Then, both with the truncated horizon approach with $s_n \in \tilde{S}$ and $a_n = 0$ for all n, and with the regenerative approach, $\theta_n \to \theta^*$ with probability one.*

If the IPA accumulator a_n is not reset to 0 between iterations, IPA has the stronger property, for this particular example and under mild additional conditions, that even when using a truncated horizon t_n that is constant with n, SA converges *weakly* to the optimizer. A proof is given in [14], based on a theorem of Kushner and Shwartz [11]. In the regenerative case, SA does not converge to the optimum in general if t_n does not converge to infinity.

4. Numerical Experiments with an M/M/1 Queue

Consider an M/M/1 queue with arrival rate $\lambda = 1$ and mean service time θ. One has $B_\theta(\zeta) = 1 - e^{-\zeta/\theta}$. Let $\Theta = [.01, .95]$ and $C(\theta) = 1/\theta$. Then, $w(\theta) = \theta/(1 - \theta)$ and $\theta^* = 0.5$. Assumptions 1 and 2 are easily verified.

We performed the following experiments. For each *variant*, i.e. each way of combining SA with a specific derivative estimator, we made 10 simulation runs, each yielding an estimation of θ^*. The 10 initial parameter values were randomly chosen, uniformly over $[.01, .95]$, and the initial state was s_0 (an empty system). Across the algorithms, we used common random numbers and the same set of initial parameter values. This means that the different entries of Table 1 are strongly correlated. Each run was stopped after 10^6 ends of service. The final state of each simulation subrun was taken as the initial state for the next one, except when stated otherwise. For FDC, the initial state s_{n+1} was the final state of the subrun at iteration n with parameter value the closest to θ_{n+1}.

For each variant, we computed the empirical mean $\bar{\theta}$, standard deviation s_d and standard error s_e of the N retained parameter values. If y_i denotes the retained parameter value for run i (i.e. the last value of θ_n), the above quantities are defined by

$$\bar{\theta} = \frac{1}{N} \sum_{i=1}^{N} y_i; \qquad s_d^2 = \frac{1}{N-1} \sum_{i=1}^{N} (y_i - \bar{\theta})^2; \qquad s_e^2 = \frac{1}{N} \sum_{i=1}^{N} (y_i - \theta^*)^2. \tag{20}$$

We also computed 95% confidence intervals I_θ on the expectation of $\bar{\theta}$, assuming that $\sqrt{N}(\bar{\theta} - E(\bar{\theta}))/s_d$ follows a Student distribution with $N - 1$ degrees of freedom. The results appear in the third column of Table 1.

In the Table, LRR refers to the regenerative version of LR given in (17), while IPAR refers to the regenerative version of IPA. The symbol -0 means that the state was reset to $s_0 = 0$ at the beginning of each iteration. The symbol -Z following IPA means that the IPA accumulator was reset to 0 between iterations. LR-D means the "decomposed" version of LR given by (18). LR-C [LR-DC] means LR [LR-D] in which $h_t(\theta, s, \omega)$ was replaced by $h_t(\theta, s, \omega) - 1$. This does not change the expectation of $\psi_t(\theta, s, \omega)$, but reduces its variance from $O(t)$ to $O(1)$ at $\theta = \theta^*$, because $w(\theta^*) = 1$ (see [13]). In all cases, we had $\gamma_n = 1/n$. We took $c_n = 0.1n^{-1/6}$ for FD and $c_n = 0.1n^{-1/5}$ for FDC. For FDC, we also tried $c_n = 0.001n^{-2}$, which is denoted by FDC-NN.

We see that IPA performs well, even when t_n is fixed at a small constant. IPA-Z, IPAR, FDC, and FDC-NN, with a linearly increasing t_n, are approximately as good. When t_n is fixed to a small constant, convergence is also quick with FDC, IPA-Z, or IPAR (small s_d), but the standard error s_e is very large, which indicates that convergence is not to the right value. Even for $t_n = 100$, the bias is still quite apparent for FDC. The LR methods in general have trouble due to their large associated variance, and large bias when t_n grows slowly. LR with $t_n = n^p$ has large variance for large p, and for small p, the bias goes down much too slowly compared to the variance. As a result, the confidence interval I_θ, based on the N final values of θ_n, is very likely not to cover θ^*. This is what happens, for instance, with $p = 1/3$. Among the truncated-horizon variants, LR-C and LR-CD provide significant improvements over LR. The LR variant that gives the best results here is LRR (regenerative) with t_n increasing linearly. With $t_n = n^{1/2}$, both LRR and FDC have the same bias problem as described above: the bias goes down too slowly and I_θ does not contain θ^*. Nevertheless, they converge (slowly) to the right answer (we verified it empirically with longer simulation runs).

Independent sets of experiments were also performed with $\bar{T} = 10^5$ and the results were quite similar to the ones given here [4]. We also made experiments with $C(\theta) = 1/(250\theta)$ (for which $\theta^* = 1/6$) and $C(\theta) = 250$ (for which $\theta^* = 5/6$). The results appear in the last two columns of Table 1.

	T_n	$C_1 = 1$ ($\theta^* = 1/2$)		$C_1 = 1/25$ ($\theta^* = 1/6$)		$C_1 = 25$ ($\theta^* = 5/6$)	
		s_d	s_e	s_d	s_e	s_d	s_e
FD	n	.00979	.00967				
FD	$100 + n$	.01075	.01044				
FDC	5	.00149	.15343 ◁				
FDC	100	.00340	.00721 ◁				
FDC	n	.00193	.00184	.00030	.00030	.02354	.02234
FDC	$100 + n$	.00204	.00198	.00027	.00029	.02875	.02824
FDC-0	n	.00243	.00231	.00039	.00037	.03019	.02867
FDC-NN	n	.00203	.00196	.00031	.00031	.02270	.02177
FDC	$n^{1/2}$	.00181	.00684 ◁				
IPA	1	.00227	.00217				
IPA	10	.00227	.00216	.00053	.00051	.02402	.02575
IPA	100	.00229	.00219				
IPA	n	.00195	.00185	.00046	.00044	.03208	.03416
IPA	$100 + n$	.00203	.00193	.00046	.00043	.02685	.02849
IPA-Z	10	.00169	.07365 ◁				
IPA-Z	n	.00192	.00189	.00046	.00044	.02449	.02597
IPA-0	n	.00246	.00233	.00042	.00040	.01721	.01956
IPAR	5	.00228	.06175 ◁				
IPAR	n	.00200	.00197	.00046	.00044	.02981	.03110
LR	$n^{1/3}$	.01221	.02062 ◁				
LR	$n^{1/2}$	.03012	.02876	.02454	.02355	.04473	.05214
LR	$n^{2/3}$	.07494	.07115				
LR-C	$n^{1/2}$	.00772	.00749	.00221	.00291 ◁	.03433	.04864 ◁
LR-C0	$n^{1/2}$	.00709	.00725				
LR-D	$n^{1/2}$	.01502	.01658				
LR-CD	$n^{1/2}$	.00533	.00615	.00175	.00176	.03000	.05141 ◁
LR-CD	$n^{2/3}$	.00706	.00688	.00264	.00255	.04893	.04857
LRR	n	.00447	.00453	.00124	.00118	.07608	.07446
LRR	$n^{1/2}$	.00443	.01775 ◁				

Table 1: Some experimental results for 10 times 10^6 customers.
For the values marked with ◁, the 95% confidence interval does not contain θ^*.

ACKNOWLEDGMENTS

The work of the first author was supported by NSERC-Canada grant # A5463 and FCAR-Québec grant # EQ2831. The third author's work was supported by the IBM corporation under SUR-SST contract 12480042 and by the U.S. Army Research Office under contract DAAL-03-88-K-0063.

REFERENCES

[1] Aleksandrov, V. M., V. I. Sysoyev and V. V. Shemeneva, "Stochastic Optimization", *Engineering Cybernetics*, **5** (1968), 11–16.

[2] Asmussen, S., *Applied Probability and Queues*, Wiley, 1987.

[3] Bratley, P., B. L. Fox, and L. E. Schrage, *A Guide to Simulation*, Springer-Verlag, New York, Second Edition, 1987.

[4] Giroux, N. "Optimisation Stochastique de Type Monte Carlo", Mémoire de maîtrise, dépt. d'informatique, Univ. Laval, jan. 1989.

[5] Glynn, P. W. "Likelihood Ratio Gradient Estimation: an Overview", *Proceedings of the Winter Simulation Conference 1987*, IEEE Press (1987), 366–375.

[6] Glynn, P. W. "Optimization of Stochastic Systems Via Simulation", *Proceedings of the Winter Simulation Conference 1989*, IEEE Press (1989), 90–105.

[7] Glynn, P. W. "Likelihood Ratio Gradient Estimation for Stochastic Systems", *Communications of the ACM*, **33**, 10 (1990), 75–84.

[8] Heidelberger, P., Cao, X.-R., Zazanis, M. A. and Suri, R., "Convergence Properties of Infinitesimal Perturbation Analysis Estimates", *Management Science*, **34**, 11 (1989), 1281–1302.

[9] Ho, Y.-C., "Performance Evaluation and Perturbation Analysis of Discrete Event Dynamic Systems", *IEEE Transactions of Automatic Control*, **AC-32**, 7 (1987), 563–572.

[10] Kushner, H. J. and Clark, D. S., *Stochastic Approximation Methods for Constrained and Unconstrained Systems*, Springer-Verlag, Applied Math. Sciences, vol. 26, 1978.

[11] Kushner, H. J. and Shwartz, A., "An Invariant Measure Approach to the Convergence of Stochastic Approximations with State Dependent Noise", *SIAM J. on Control and Optim.*, **22**, 1 (1984), 13–24.

[12] L'Ecuyer, P., "A Unified View of the IPA, SF, and LR Gradient Estimation Techniques", *Management Science*, **36**, 11 (1990), 1364–1383.

[13] L'Ecuyer, P. and Glynn, P. W., "A Control Variate Scheme for Likelihood Ratio Gradient Estimation", In preparation (1990).

[14] L'Ecuyer, P., Giroux, N., and Glynn, P. W., "Stochastic Optimization by Simulation: Convergence Proofs and Experimental Results for the GI/G/1 Queue", manuscript, 1990.

[15] Meketon, M. S., "Optimization in Simulation: a Survey of Recent Results", *Proceedings of the Winter Simulation Conference 1987*, IEEE Press (1987), 58–67.

[16] Pflug, G. Ch., "On-line Optimization of Simulated Markovian Processes", *Math. of Oper. Res.*, **15**, 3 (1990), 381–395.

[17] Reiman, M. I. and Weiss, A., "Sensitivity Analysis for Simulation via Likelihood Ratios", *Operations Research*, **37**, 5 (1989), 830–844.

[18] Rubinstein, R. Y., *Monte-Carlo Optimization, Simulation and Sensitivity of Queueing Networks*, Wiley, 1986.

[19] Rubinstein, R. Y., "The Score Function Approach for Sensitivity Analysis of Computer Simulation Models", *Math. and Computers in Simulation*, **28** (1986), 351–379.

[20] Rubinstein, R. Y., "Sensitivity Analysis and Performance Extrapolation for Computer Simulation Models", *Operations Research*, **37**, 1 (1989), 72–81.

[21] Suri, R., "Infinitesimal Perturbation Analysis of General Discrete Event Dynamic Systems", *J. of the ACM*, **34**, 3 (1987), 686–717.

[22] Suri, R., "Perturbation Analysis: The State of the Art and Research Issues Explained via the GI/G/1 Queue", *Proceedings of the IEEE*, **77**, 1 (1989), 114–137.

[23] Suri, R. and Leung, Y. T., "Single Run Optimization of Discrete Event Simulations—An Empirical Study Using the M/M/1 Queue", *IIE Transactions*, **21**, 1 (1989), 35–49.

OPTIMIZATION OF STOCHASTIC DISCRETE EVENT DYNAMIC SYSTEMS: A SURVEY OF SOME RECENT RESULTS

Alexei A.Gaivoronski

V.Glushkov Institute of Cybernetics, Kiev, USSR

1. Introduction.

Models of discrete event dynamic systems (DEDS) include finite state machines [24], Petri nets [35], finitely recursive processes [26], communicating sequential processes [18], queuing models [41] among others. They become increasingly popular due to important applications in manufacturing systems, communication networks, computer systems. We would consider here a system which evolution or sample path consists of the sequence

$$Y(x,\omega)=\Big\{(t_0,z_0),(t_1,z_1),\ldots,(t_s,z_s),\ldots.\Big\}, \quad t_i=t_i(x,\omega), \quad z_i=z_i(x,\omega)$$

where $z_i(x,\omega)\in W$ is the state of the system during the time interval $t_i(x,\omega)\leq t<t_{i+1}(x,\omega)$, $x\in X\subseteq\mathbb{R}^n$ is the set of control parameters and $\omega\in\Omega$ is an element of some probability space $(\Omega,\mathbb{B},\mathbb{P})$. Particular rules which govern transitions between states at time moments t_i can be specified in the framework of one of the models mentioned above. For describing the time behavior the generalized semi- Markov processes proved to be useful [47]. We consider the following optimization problem:

$$\text{minimize} \quad F(x) = \mathbb{E}f(Y(x,\omega),x,\omega) \tag{1}$$

subject to constraints

$$x\in X \tag{2}$$

where the sample objective function $f(Y(x,\omega),x,\omega)$ depend on control parameters both directly and indirectly through the sample path $Y(x,\omega)$. This function may incorporate the system performance indices, like throughput, utilization of important equipment, queue lengths etc. (usually through dependence on sample path) and cost/benefit considerations (usually through direct dependence on x).

The problem (1)-(2) is a stochastic programming problem, see [5,8,28,30,46]. Usually there is no analytic expression for F(x) and it is only possible to simulate DEDS, observe some portion of the sample path $Y(x,\omega)$, make inference about $f(Y(x,\omega),x,\omega)$ and on the basis of such simulations obtain statistical estimate of the value of

F(x) or its gradient. These estimates are used in the stochastic quasigradient projection method [5]

$$x^{s+1}=\pi_X(x^s-\rho_s\xi^s) \tag{3}$$

to obtain approximation of the optimal solution. Here π_X is the projection operator on the set X, $\rho_s\geq 0$ is the step size and ξ^s is a statistical estimate of the gradient of the function F(x) at the point x^s called stochastic quasigradient:

$$\mathbb{E}(\xi^s|x^0,\ldots,x^s)=F(x^s)+a_s \tag{4}$$

where a_s is some diminishing term. For some particular choices of ρ_s, ξ^s some variants of this method were discovered in statistics where the usual term for them is stochastic approximation [29,30]. Variety of strategies for selection of the step size ρ_s and step direction ξ^s are discussed in [8] together with convergence results, advanced computer implementation is described in [10]. Various similar algorithms for concurrent simulation and optimization were considered in [6,14,23,36,45]

In order to be applied to the problem (1)-(2) the method (3)- (4) needs considerable adaptation and further development due to the following peculiarities of the problem.

1. The function $f(Y(x,\omega),x,\omega)$ is defined on the sample path $Y(x,\omega)$ which is often infinite, for instance when it is necessary to optimize the stationary behavior. Only finite portions of $Y(x,\omega)$ are observable and therefore in many cases we have only more or less accurate approximations to the values of $f(Y(x,\omega),x,\omega)$, or only biased estimates of $F_x(x)$, where such estimates are available. One exception is regenerative processes [4], where sufficient information on the stationary behavior of the system is contained in appropriately defined finite portion of the sample path. This case, however, is quite special and, moreover, the required portion (regenerative cycle) may still be too long. The important issue here is how to subdivide the sample path $Y(x,\omega)$ into portions $Y_s(x,\omega)$, s=1,2,...

$$Y(x,\omega)=Y_1(x,\omega)\cup Y_2(x,\omega)\cup\cdots\cup Y_s(x,\omega)\cup\cdots \tag{5}$$

in such a way that the reasonable estimate of the step direction ξ^s can be obtained on the basis of $Y_s(x,\omega)$ only, which then will be used in the algorithm (3)-(4). In this process of concurrent simulation and optimization for each portion $Y_s(x,\omega)$ there will be its value x^s of control variables x obtained by (3) and the composite sample path $Y(x,\omega)$ would depend actually on the whole sequence x^s, s=0,1,... If subdivision (5) is appropriately organized and the process is ergodic, then the solution of the problem (1)-(2) will be obtained using only one sample path. Some results in this direction are contained in

[6,36].

2. The problem (1)-(2) has a lot of structure which is contained in the rules which define the evolution of the sample path $Y(x,\omega)$. How to use this structure in order to obtain efficient estimates ξ^s of the gradient $F_x(x^s)$ and minimize the amount of simulation needed? There is considerable activity going on in this direction and notable results has been obtained which are surveyed in the section 2. However many important issues are still unresolved. In particular, some of the existing differentiation schemes fail to cope with inherent discontinuities of the problem, while others provide estimates which variance grows rapidly with the number of simulation steps.

3. A lot of information about the value of $F_x(x)$ at the current point x^s is contained in simulations performed at the previous steps of (3). This is due to the fact that usually the value ρ_s is small. How to use this information in order to improve the estimate ξ^s and cut the total amount of simulation? One possible approach is presented in the section 3 of this paper. In the course of optimization the function $F(x)$ is approximated using observations of the sample performance function $f(Y(x,\omega),x,\omega)$ at the previous point and the estimate ξ^s of $F_x(x^s)$ is computed using this approximation. Therefore this approach is applicable when differentiation schemes fail and present considerable improvement over conventional finite differences.

2. Obtaining statistical estimates of the gradient (perturbation analysis).

In this section we consider various approaches for computing statistical estimates of the gradient $F_x(x)$ of the function defined by averaging over the set of sample paths of discrete event system and survey some of the difficulties encountered by several such schemes. This is done for the purpose of putting in the right context a new differentiation scheme presented at the end of this section and approach presented in the next section. It is considered only how various methods deal with inherent discontinuities of the sample paths and with increase in the length of the evolution steps, as well as amount of simulation needed. Much more comprehensive survey is presented in [44].

Let us start by examining in more detail the structure of the problem (1)-(2). We have the following representation:

$$F(x)=Ef(Y(x,\omega),x,\omega)=\int_\Omega f(Y(x,\omega),x,\omega)H(d\omega) \tag{6}$$

where H is some measure defined on the set Ω. For the sake of clarity let us assume in this section that the sample path $Y(x,\omega)$ consists of finite number of steps:

$$Y(x,\omega)=\Big\{(t_0,z_0),(t_1,z_1),\ldots,(t_k,z_k)\Big\}\ t_i=t_i(x,\omega),\ z_i=z_i(x,\omega)$$

where k is fixed. The cases when k depends on ω or is infinite are treated similarly, but with more involved technicalities. Both function $f(Y(x,\omega),x,\omega)$ and measure H usually have a special structure.

1. It is recognized feature of discrete event dynamic systems that their sample path $Y(x,\omega)$, and in particular the sequence of states $z_i(x,\omega)$, usually depends discontinuously on (x,ω) [19]. In simulation literature this phenomenon is referred to as "event order changes" [3]. As the result of this, the function $f(Y(x,\omega),x,\omega)$ would as a rule be discontinuous with respect to (x,ω), but in the most situations this will not prevent the averaged function $F(x)$ from being continuous and differentiable. Example described in the Appendix A of this paper shows that discontinuities appear even in the very simple systems. Usually to each sequence $Z(x,\omega)$:

$$Z=Z(x,\omega)=\Big\{z_0(x,\omega),z_1(x,\omega),\ldots,z_k(x,\omega)\Big\} \tag{7}$$

corresponds subset $\Theta(Z)\subseteq X\times\Omega$ such that if $(x_1,\omega_1)\in\Theta(Z')$, $(x_2,\omega_2)\in\Theta(Z')$ then the simulation of the system with values of (x,ω) equal to (x_1,ω_1) and (x_2,ω_2) would produce the same sequence of states Z' and only times $t_i(x,\omega)$ when the system changes its state would differ. The function $f(Y(x,\omega),x,\omega)$ would be continuous on the sets $\Theta(Z)$ and discontinuities will occur on the boundaries of those sets. Let us denote by Z the set of all sequences $Z(x,\omega)$. In order not to overburden exposition with secondary details we consider the case when K is finite, although analysis which follows is valid for infinite k too. Let us take

$$\Theta(x,Z)=\Big\{\omega:\ (x,\omega)\in X\times\Omega\Big\}$$

Then the sets $\Theta(Z)$ and $\Theta(x,Z)$ constitute partition of $X\times\Omega$ and Ω respectively, i.e.

$$\Theta(Z_1)\cap\Theta(Z_2)=\varnothing,\ \ \Theta(x,Z_1)\cap\Theta(x,Z_2)=\varnothing,\ \ Z_1\neq Z_2,\ \bigcup_{Z\in Z}\Theta(Z)=X\times\Omega,$$

$$\bigcup_{Z\in Z}\Theta(x,Z)=\Omega \tag{8}$$

2. Since the evolution of discrete event system consist of a sequence of steps which gradually unfold in time, the set Ω can be often represented as the product of sets Ω_i, each corresponding to one particular step of the system evolution. To each of those sets correspond probability measure $H_i(dw_i)$:

$$\omega=(\omega_1,\ldots,\omega_k), \quad \Omega=\underset{i=1}{\overset{k}{\times}}\Omega_i, \quad H(d\omega)=\prod_{i=1}^{k}H_i(dw_i), \tag{9}$$

Here the measure $H_i(d\omega_i)$ would usually depend on the evolution of the system during previous steps $j<i$.

Both these features are crucial for obtaining statistical estimates of the gradient and represent the source of some of the difficulties.

Taking into account (8),(9) we obtain the following representation for the objective function of the problem (1)-(2):

$$F(x)=\sum_{Z\in Z}\int_{\Theta(x,Z)}f(Y(x,\omega),x,\omega)\prod_{i=1}^{k}H_i(dw_i) \tag{10}$$

where each term is an integral of continuous function of x taken over the set which also depends on x. Now we can proceed with survey of approaches to obtain a statistical estimate of the gradient of such function.

Finite differences. The simplest way is to use forward differences

$$\xi_{FD}^s=\sum_{i=1}^{n}\frac{f(Y(x^s+\delta_s e_i,\omega^{is}),x^s+\delta_s e_i,\omega^{is})-f(Y(x^s,\omega^{0s}),x^s,\omega^{0s})}{\delta_s}e_i \tag{11}$$

or similar expression for central finite differences. Here e_i are unit vectors of $\mathbb{R}^n$, ω^{is}, $i=0:n$ are independent observations of ω, each corresponds to the separate run of the model. This approach has two serious shortcomings:

- it requires at least $n+1$ simulation runs which grows to $2n$ for central finite differences;

- the variance of the estimate (11) approaches infinity while $\delta_s\rightarrow0$ since for independent observations

$$\mathbb{E}(\|\xi_{FD}^s-\mathbb{E}\xi_{FD}^s\|^2|x^0,\ldots,x^s)=\frac{1}{\delta_s^2}\sum_{i=1}^{n}(C_{si}+C_{s0}) \tag{12}$$

where

$$C_{si}=\mathbb{E}((f(Y(x^s+\delta_s e_i,\omega^{is}),x^s+\delta_s e_i,\omega^{is})-F(x^s+\delta_s e_i))^2|x^0,\ldots,x^s),$$

$$C_{s0}=\mathbb{E}((f(Y(x^s,\omega^{0s}),x^s,\omega^{0s})-F(x^s))^2|x^0,\ldots,x^s)$$

on the other hand, taking large values of δ_s would decrease variance, but lead to significant systematic error.

The simulation effort can be reduced considerably and the problem with increasing variance could be alleviated if all observations of $f(Y(x,\omega),x,\omega)$ needed to compute (11) could be made for the same value of random parameters ω. This, however, can not be done straightforwardly since here discontinuity comes into play. To see this, let us derive a formula for the gradient of the function F(x) from (10).

$$F_x(x) = \sum_{i=1}^{n} \left(\sum_{Z \in Z} \lim_{\Delta \to 0} \frac{1}{\Delta} \left(\int_{\Theta(x+\Delta e_i, Z)} f(Y(x+\Delta e_i, \omega), x+\Delta_i e_i, \omega) \prod_{i=1}^{k} H_i(dw_i) - \right. \right.$$

$$\left. \left. \int_{\Theta(x, Z)} f(Y(x, \omega), x, \omega) \prod_{i=1}^{k} H_i(dw_i) \right) \right) e_i \tag{13}$$

where e_i is a unit vector of $\mathbb{R}^n$. Suppose that for each given $Z \in Z$ the function $f(Y(x, \omega), x, \omega)$ is differentiable with respect to X. Then under fairly mild conditions (13) would be equal to the sum of two terms:

$$F_x(x) = G_1(x) + G_2(x) \tag{14}$$

where

$$G_1(x) = \sum_{i=1}^{n} \left(\sum_{Z \in Z} \int_{\Theta(x, Z)} \lim_{\Delta \to 0} \frac{1}{\Delta} (f(Y(x+\Delta e_i, \omega), x+\Delta_i e_i, \omega) - \right.$$

$$\left. f(Y(x, \omega), x, \omega)) \prod_{i=1}^{k} H_i(dw_i) \right) e_i =$$

$$\sum_{Z \in Z} \int_{\Theta(x, Z)} \frac{d}{dx} f(Y(x, \omega), x, \omega) \prod_{i=1}^{k} H_i(dw_i) \tag{15}$$

$$G_2(x) = \sum_{i=1}^{n} \left(\sum_{Z \in Z} \lim_{\Delta \to 0} \frac{1}{\Delta} \left(\int_{\Theta(x+\Delta e_i, Z)} f(Y(x, \omega), x, \omega) \prod_{i=1}^{k} H_i(dw_i) - \right. \right.$$

$$\left. \left. \int_{\Theta(x, Z)} f(Y(x, \omega), x, \omega) \prod_{i=1}^{k} H_i(dw_i) \right) \right) e_i \tag{16}$$

The term $G_2(x)$ takes account of discontinuities of the function $f(Y(x, \omega), x, \omega)$ on the boundaries of the sets $\Theta(x, Z)$ and vanishes if this function is continuous.

From (14)-(16) follows that if we make all function evaluations in the finite difference formula (11) for the same value of ω, this would be equivalent to estimating the term $G_1(x)$ and neglecting the term $G_2(x)$. For most systems such an estimate would have a large bias. There is a way around this problem. It is possible to smooth the function $f(Y(x, \omega), x, \omega)$ by introducing a specially selected noise in the control variables x. This would eliminate the term $G_2(x)$ and allow use of the same values of ω, but would also introduce a bias which depends on the level of noise. For more details see [6] where it was also proposed to use random directions for computing finite differences in order to decrease the amount of simulation.

Among positive features of the finite differences is that this technique is not sensitive to the number of simulation steps, on the contrary, in the ergodic case increase in the number of steps k would lead to more accurate estimates.

Special technique called Extended Perturbation Analysis was introduced in [20] where the ergodic properties of the process are used in order to reduce the simulation effort considerably.

Infinitesimal Perturbation Analysis (IPA). This is the pioneering attempt to depart from the "brute force" approach of finite differences and exploit instead the structure of discrete event systems in order to obtain statistical estimates of the gradient. In fact, current interest in perturbation analysis of DEDS, including this paper, is very much due to efforts of Prof. Y.C.Ho and his colleagues [15,17,19-22,43,44], for more extensive bibliography see [22,44]. In its original form the IPA estimate for the gradient of the function (10) is:

$$\xi_{IPA}^{s} = \frac{d}{dx} \, f(Y(x,\omega^{s}),x,\omega^{s}) \tag{17}$$

where ω^{s} corresponds to one simulation run. Using the structure of DEDS it is possible to organize the computation of ξ_{IPA}^{s} very efficiently and compute all components of the gradient during a single simulation run. Often both reduction in amount of simulation and accuracy would reach order of magnitude compared with conventional finite differences (11).

However, comparing (17) with (14)-(16) we see that the scope of application of the IPA estimate is limited since it would estimate only $G_1(x)$ and can not cope with the term $G_2(x)$ which appears due to discontinuities. Hence, the IPA estimate would be biased in each case when discontinuities ("event order changes") are significant. Still, the number of cases when the IPA would be unbiased is somewhat larger then (14)-(16) would suggest. In some ergodic cases it appears that if the number of evolution steps k during one simulation run tends to infinity then the terms which constitute the sum in (16) would asymptotically cancel each other out. Sometimes this happens even for the finite k. These cases include some important models from queuing theory, for survey see [17]. However these are quite special cases.

Various extensions of IPA were designed in order to overcome this difficulty. In [20] the IPA estimates were combined with the finite differences. In [15] the smoothed perturbation analysis was introduced. The idea is to select some random variable $\eta=\eta(Y(x,\omega),x,\omega)$ which depend on the sample path $Y(x,\omega)$ and such that the conditional expectation $\bar{F}(x,\eta)=E(f(Y(x,\omega),x,\omega)|\eta)$ is smooth. This operation would cancel discontinuities term $G_2(x)$ in (14) which would make possible to apply the IPA to $\bar{F}(x,\eta)$. However the selection of such smoothing variable η for a wide class of DEDS is difficult and not yet resolved problem. Other recent extensions of IPA include [12,16,25]. The IPA estimates when they are unbiased possess the following attractive feature. If the system is ergodic their accuracy often grows with the number of evolution steps k.

Likelihood ratio/score function method (LR/SF). This differentiation technique goes back to [2], was developed for various stochastic systems in [13,14,39,40], in particular for systems described by semi-Markov processes in [7], for general class of stochastic Petri Nets in [1]. In order to describe those estimates in our context let us specify more precisely the structure of the function (1),(10).

Suppose that the sample performance function f depends on (x,ω) only through the sample path $Y(x,\omega)$:

$$f(Y(x,\omega),x,\omega)=f(Y(x,\omega)) \qquad (18)$$

and exist a collection of independent random variables $V_k(x,\omega)$:

$$V_k(x,\omega)=\left\{v_1(x,\omega),\ldots,v_1(x,\omega),\ 1=1(k)\right\}$$

with distribution functions $P_i(x,v_i)$ such that the sample path can be expressed as a function of these variables:

$$Y(x,\omega)=Y(V_k)=Y(V_k(x,\omega)) \qquad (19)$$

If conditions (18)-(19) hold then it is possible to substitute in (10) integration variables ω by variables v and obtain the following expression:

$$F(x)=\sum_{Z\in Z}\int_{\theta(Z)} f(Y(V_k))\prod_{i=1}^{1(k)}dP_i(x,v_i) = \int_V f(Y(V_k))\prod_{i=1}^{1(k)}dP_i(x,v_i) \qquad (20)$$

where V is the set of values of $V_k(x,\omega)$, which is usually $\mathbb{R}^{1(k)}$, $\theta(Z)$ is the image of the set $\theta(x,Z)$ in the set V. Although the function $f(Y(V_k))$ remain discontinuous on the boundaries of the sets $\theta(Z)$, these sets cease to depend on the control variables x. This means that the discontinuities term $G_2(x)$ in (16) vanishes after such transformation and the whole issue of discontinuities is substituted by the question of differentiability of the measure which correspond to the joint distribution of V_k [37]. The required set of variables V_k is usually easy to find, these are usually durations of events and state transformations, they can have continuous as well as discrete distributions. All components of the gradient can be estimated during single simulation run with very moderate overhead using specially selected sampling distributions $\bar{P}_i(x,v_i)$ which may or may not coincide with $P_i(x,v_i)$. In case if $P_i(x,v_i)$, $\bar{P}_i(x,v_i)$ has continuous densities $p_i(x,v_i)$, $\bar{p}_i(x,v_i)$ and $p_i(x,v_i)$ are differentiable with respect to x the LR/SF estimate looks as follows:

$$\xi_{LR/SF}^s = f(Y(V_k^s))\tau_{1(k)} \qquad (21)$$

where the vector multiplier $\tau_{1(k)}$ is computed according to the simple recursive scheme:

$$\tau_0=0, \quad \nu_0=1, \quad \nu_{i+1}=\frac{p_i(x,v_i)}{\bar{p}_i(x,v_i)}\,\nu_i,$$

$$\tau_{i+1}=\frac{p_i(x,v_i)}{\bar{p}_i(x,v_i)}\,\tau_i+\frac{\frac{d}{dx}p_i(x,v_i)}{\bar{p}_i(x,v_i)}\,\nu_i \tag{22}$$

in which the simulated values of v_i are substituted. Similar schemes hold in the case of discrete distributions with probabilities differentiable with respect to x [1,7].

At the first glance this approach looks like the ultimate solution to the problem of discontinuities. However it has limitations of its own, namely its variance grows with the number k of evolution steps. To see this, let us consider the estimate (21) for the simplest case when $p_i(x,v_i)=\bar{p}_i(x,v_i)$. From (22) follows that

$$\mathbb{E}\tau_{1(k)}=0, \quad \mathbb{E}(\tau_{1(k)}-\mathbb{E}\tau_{1(k)})^2=\sum_{i=1}^{1(k)}\int\frac{p_{ix}^2(x,v_i)}{p_i(x,v_i)}\,dv_i \tag{23}$$

Thus, each generation of random variable would add additional positive term in the variance of the multiplier $\tau_{1(k)}$ and the variance of the estimate $\xi_{LR/SF}^s$ would grow in a similar manner. Therefore the accuracy of LR/SF estimate would deteriorate with the number of evolution steps k since l(k) is usually proportional to k. Finite differences and IPA estimates, on the contrary, may improve with k. This deterioration may be so pronounced that even for relatively simple systems the LR/SF estimate may become inferior to finite differences in terms of accuracy, for the relevant example see Appendix B.

This method is also under vigorous development now and its versions with improved performance have been introduced recently [32,42].

Weak derivatives (WD). [36-38] Similarly to the score functions approach, this method tries to put all explicit dependence on the control parameters x in the distribution function of random variables and considers the function F(x) of the form (20). It differs substantially in the way how sampling is organized. The gradient of F(x) is presented in the form:

$$F_x(x)=\frac{d}{dx}\int_V f(Y(V_k))\prod_{i=1}^{1(k)}dP_i(x,v_i)=\int_V f(Y(V_k))(\dot{P}(x,dV_k)-P(x,dV_k))$$

where the pair $(\dot{P}(x,V_k),P(x,V_k))$ is the weak derivative of the measure P(x) of the joint distribution of the variables V_k. Under mild assumptions such derivative exist and calculus of such differentiation is developed which allows to organize the sampling from measures $\dot{P}(x),P(x)$ in efficient way [37]. The weak derivative ξ_{WD}^s is obtained through such sampling. The estimation can be performed using single simulation run plus some additional sampling which accounts for two sampling measures. Similarly to LR/SF estimate this method copes perfectly with discontinuities. The estimate variance would again grow with the number of evolution steps k, but according to examples and theoretical considerations presented in [38], this growth may be considerably slower then in the case of the score functions estimate.

Augmented Perturbation Analysis (APA). [11]   This   is   a   recent differentiation scheme which addresses the issue of discontinuities directly by providing a statistical estimate $\hat{\xi}^s$ for discontinuities term $G_2(x)$ from (16). This estimate provides an unbiased gradient estimate for a general class of discrete event dynamic systems, and needs one simulation run with moderate overhead for additional analysis of the process behavior. It looks as follows:

$$\xi_{APA}^s = \xi_{IPA}^s + \hat{\xi}^s \tag{24}$$

where ξ_{IPA}^s is an IPA estimate. In this way it extends considerably the applicability of IPA. However, its variance grows with the number of sets in the partition (8), although the growth with the number of evolution steps k would be slower then in the case of the score function estimate.

Thus, each existing gradient estimation techniques has its strong and weak points and approaches which cope successfully with discontinuity problem exhibit "dual deficiency" of increasing variance in the course of simulation steps. The following question is still unresolved:

Is it possible to develop a gradient estimate which
- would be reasonably computable for general DEDS,
- would be insensitive to the simulation length (like IPA),
- would be insensitive to discontinuities (like score functions)?

3. Concurrent approximation and optimization.

We have seen from the last section that for a large class of discrete event systems development of efficient differentiation algorithms is still an open question. Those are the systems for which discontinuities (event order changes) are significant and observation of the system behavior may require from moderate to large number of steps. For such systems it may be advantageous to develop methods which are based only on observations of the sample performance function and intelligently use information obtained during previous iterations of the algorithm (3)-(4).

In this section we introduce a general approach for constructing stochastic optimization algorithms which is based on observations of the values of the objective function only. This approach combines stochastic quasigradient techniques (3)-(4) with response surface methodology, for general stochastic programming problems similar approach was considered in [9]. It is not limited to discrete event systems. However, it is particularly useful for optimization of DES when direct application of differentiation schemes is difficult. It needs considerably less simulation effort compared with other techniques which does not directly involve differentiation. We specify one new algorithm based on this approach and present a convergence theorem, its proof and results of numerical experiments are reported in [6].

Informally speaking, the idea behind proposed approach is the following. Suppose that in the course of optimization were obtained the sequence of points $x^0,...,x^s$ and the set of observations $\zeta_1,...,\zeta_s$ such that $E(\zeta_i|x^0,...,x^i)=F(x^i)$, i=0:s. These observations are used to approximate the function $F(x)$ by a function $F(s,x)$. Let $\bar{x}^s \in X$ be a point at which $F(s,x)$ attains its minimal value over the set X. Then the next approximation to the optimal solution of the problem (4) is obtained as a linear combination of x^s and $\bar{x}^s$:

$$x^{s+1}=(1-\rho_s)x^s+\rho_s\bar{x}^s$$

or, it is obtained by making a step in the direction opposite to the gradient of the approximating function:

$$x^{s+1}=\pi_X(x^s-\rho_sF_x(s,x^s))$$

After that a new observation is made, the approximation $F(s,x)$ is updated using this observation and the process continues.

Let us compare this approach with two other techniques which does not use derivatives: finite differences and response surface methods. Shortcomings of the finite differences were discussed in the previous section. Here we point out that all observations of the objective function which are made at the point x^s in order to obtain an estimate of the gradient via finite differences (13) are discarded on the next iteration when all observations are made again at the point x^{s+1}. At the same time these observations contain considerable amount of information on the value of $F_x(x^{s+1})$ since the stepsize ρ_s is usually small and $F(x)$ is continuously differentiable. The approach which we propose here use all this information, which result in estimates with smaller variance and/or smaller simulation effort since it can work with only one new observation on each iteration.

The response surface method [27,31,33,34] constructs approximation of the objective function on the basis of observations distributed over some region, then finds the minimum of this approximate function. These steps may be repeated. The novelty of the approach proposed here is that we integrate approximation and optimization into a single on-line procedure. Approximation is updated after each step using new samples made at points (or point) obtained by optimization procedure. In this way excessive sampling in regions far from a vicinity of optimum is avoided. This again results in savings of simulation effort. Of course, extensive experimentation is needed to further validate these assertions.

In fact, much has to be done to design on its basis a practical algorithm, some of the issues to be clarified are how to choose appropriate approximation criterion, how to select approximation points properly in order to insure stability of approximations, how to discard old points, etc. Some of those issues are reflected in the Algorithm 1. Related techniques were considered in [9,23].

Algorithm 1.

1. At the beginning the initial point x^1 is chosen, $\nu_0=0$, $Y^0=\emptyset$, $\Xi^0=\emptyset$ are set.

2. Suppose that prior to making iteration number s the algorithm generated the point x^s, the set of observation points $Y^{s-1}=\left\{y^i, i=1:\nu_{s-1}\right\}$, $Y^{s-1}\subseteq X$, and the set of observations $\Xi^{s-1}=\left\{\zeta_i, i=1:\nu_{s-1}\right\}$ such that $E(\zeta_i|y^i)=F(y^i)$. The following computations are performed at the iteration number s:

i. The new set of observation points $\bar{Y}^s(x^s)=\left\{y^{1s}\ldots,y^{k_s s}\right\}$ is selected, $\bar{Y}^s\subseteq X$ and observations $\zeta_1^s,\ldots,\zeta_{k_s}^s$ are made such that $E(\zeta_i^s|y^{is})=F(y^{is})$, the sets Y^s and Ξ^s are obtained:

$$\nu_s=\nu_{s-1}+k_s, \quad Y^s=\left\{y^i, \ i=1:\nu_s, \ y^i=y^{i-\nu_{s-1},s}, \ i=\nu_{s-1}+1:\nu_s\right\},$$

$$\Xi^s=\left\{\zeta^i, \ i=1:\nu_s, \ \zeta^i=\zeta^{i-\nu_{s-1},s}, \ i=\nu_{s-1}+1:\nu_s\right\}$$

ii. The weights $\alpha_s(y)$, $y\in Y^s$ are selected, these weighs are used to define approximation criterion.

iii. The values of approximation parameters a^s are defined by solving the following approximation problem:

$$\min_{a\in A} \sum_{i=1}^{\nu_s}\alpha_s(y^i)\Phi(s,\zeta^i-F(s,a,y^i)) \tag{25}$$

where $A\subseteq R^q$, $F(s,a,x)$ is some predefined class of functions which is used to approximate $F(x)$ and the function $\Phi(s,w)$ measures the closeness of fit of the approximation $F(s,a,y)$ at the point y.

    iv. The next approximation x^{s+1} to the optimal solution is obtained either by

$$x^{s+1}=(1-\rho_s)x^s+\rho_s\bar{x}^s, \quad F(s,a^s,\bar{x}^s)=\min_{x\in X}F(s,a^s,x), \quad \bar{x}^s\in X \tag{26}$$

or

$$x^{s+1}=\pi_X(x^s-\rho_s F_x(s,a^s,x^s)) \tag{27}$$

In order to specify implementable algorithm on the basis of this scheme it is necessary to choose the approximating function $F(s,a,x)$, the approximation criterion $\Phi(s,w)$, the set of observation points Y^s and weights $\alpha_s(y)$. Some of the issues concerning convergence of this method to the optimal solution of the problem (1)-(2) for particular choices of $F(s,a,x)$, $\Phi(s,w)$, Y^s, $\alpha_s(y)$, were clarified in [9]. In the remainder of this section we shall present one algorithm not covered there.

    Let us take

$$a=(b,d), \quad b\in R^1, \quad d\in R^n, \quad A=R^{n+1}, \quad F(a,x)=b+d^T(x-x^s), \quad \Phi(s,w)=w^2 \tag{28}$$

Then the problem (25) has explicit solution

$$d^s=Q^s u^s \tag{29}$$

where

$$u^s=\sum_{i=1}^{\nu_s}\alpha_s(y^i)\left(\zeta_i-\frac{1}{\sigma_s}\sum_{j=1}^{\nu_s}\alpha_s(y^j)\zeta_j\right)(y^i-x^s), \quad \sigma_s=\sum_{i=1}^{\nu_s}\alpha_s(y^i)$$

$$Q^s=\left(\sum_{i=1}^{\nu_s}\alpha_s(y^i)(y^i-x^s)\left[(y^i-x^s)^T-\frac{1}{\sigma_s}\sum_{j=1}^{\nu_s}\alpha_s(y^j)(y^j-x^s)^T\right]\right)^{-1}$$

    Let us specify now the rule for selection of observation points. Here we would consider the case when only one observation point is added on each iteration, in order to minimize simulation requirements:

$$\bar{Y}^s=\{y^{1s}\}, \quad Y^s=\left\{y^1,\ldots,y^s\right\}, \quad y^s=x^s+r_s v^s \tag{30}$$

where v^s are independent random vectors with zero mean. Introduction of the term $r_s v^s$ is necessary in order to stabilize approximation process.

    Finally, let us specify the rule for choosing approximation weights:

$$\alpha_s(y^i)=\alpha_{is}=\begin{cases}(1-\beta_s)\alpha_{i,s-1} & \text{if } i<s \\ \beta_s & \text{if } i=s\end{cases} \tag{31}$$

where $\beta_s\leq 1$, $\beta_1=1$. It is possible to represent (29)-(31) in recursive form in order to avoid matrix inversion on each iteration [6].

    The iterations of the algorithm proceed as follows:

$$x^{s+1}=\pi_X(x^s-\rho_s\gamma_s d^s), \quad \gamma_s=\begin{cases}C_0/\|d^s\| & \text{if } \|d^s\|\geq C_0 \\ 1 & \text{otherwise}\end{cases} \tag{32}$$

The following theorem confirms convergence of the algorithm (29)-(32). By $\mathbb{B}_s$ will be denoted the σ-field defined by $x^0,\ldots,x^s$.

Theorem 1. Suppose that the following conditions are satisfied:

1. The set $X \subset \mathbb{R}^n$ is convex and compact

2. The function $F(x)$ is convex continuously differentiable and $F_x(x)$ satisfy the Lipshitz condition on X.

3. $E(v^s v^{s^T} | \mathbb{B}_s) = E v^s v^{s^T} = V > 0$, $\quad E(v^s | \mathbb{B}_s) = 0$, $\quad \|v^s\| < C < \infty$,

$\quad E(\zeta^s - F(y^s) | \mathbb{B}_s, v^s) = 0$, $\quad E((\zeta^s - F(y^s))^2 | \mathbb{B}_s, v^s) < C < \infty$.

4. $\beta_s \geq 0$, $\quad \sum_{i=1}^{\infty} \beta_i = \infty$, $\quad (1-\beta_s)\dfrac{r_{s-1}^2}{r_s^2} = 1 - \beta_{1s}$, $\quad \dfrac{\beta_s}{\beta_{1s}} \to 1$, $\quad r_s \to 0$, $\quad r_{s-1} \geq r_s$,

$\dfrac{r_{s-1}}{r_s} \to 1$, $\quad \dfrac{\rho_{s-1}}{\beta_s r_s^2} \to 0$, $\quad \sum_{i=1}^{\infty} \dfrac{\beta_i^2}{r_i^2} < \infty$, $\quad \sum_{i=1}^{\infty} \dfrac{\rho_{i-1}^2}{r_i^4} < \infty$, $\quad \dfrac{1}{\beta_s}\left| \dfrac{\rho_{s-2}\beta_s}{\rho_{s-1}\beta_{s-1}} - 1 \right| \to 0$,

$$\rho_s \geq 0, \quad \sum_{i=1}^{\infty} \rho_i = \infty$$

Then sequence x^s has accumulation points and all such points belong to the set X^* of solutions of the problem (1)-(2). $\square$

More discussion, proof of this theorem and results of numerical experiments are presented in [6]. If the sequences r_s, β_s and ρ_s behave asymptotically like $s^{-r}, s^{-\beta}$ and $s^{-\rho}$ then the condition 4 is satisfied for

$$\rho \leq 1, \quad \beta < 1, \quad \rho - \beta - 2r > 0, \quad 2\beta - 2r > 1, \quad 2\rho - 4r > 1$$

for instance for $\rho = 1$, $\beta = 0.7$, $r = 0.14$.

APPENDIX A. EXAMPLE OF DISCRETE EVENT SYSTEM WITH DISCONTINUITIES

Suppose that the manufacturing system contains two machines M_1, M_2 and the buffer B. The buffer contains items which should be processed consecutively by both machines (Figure 1).

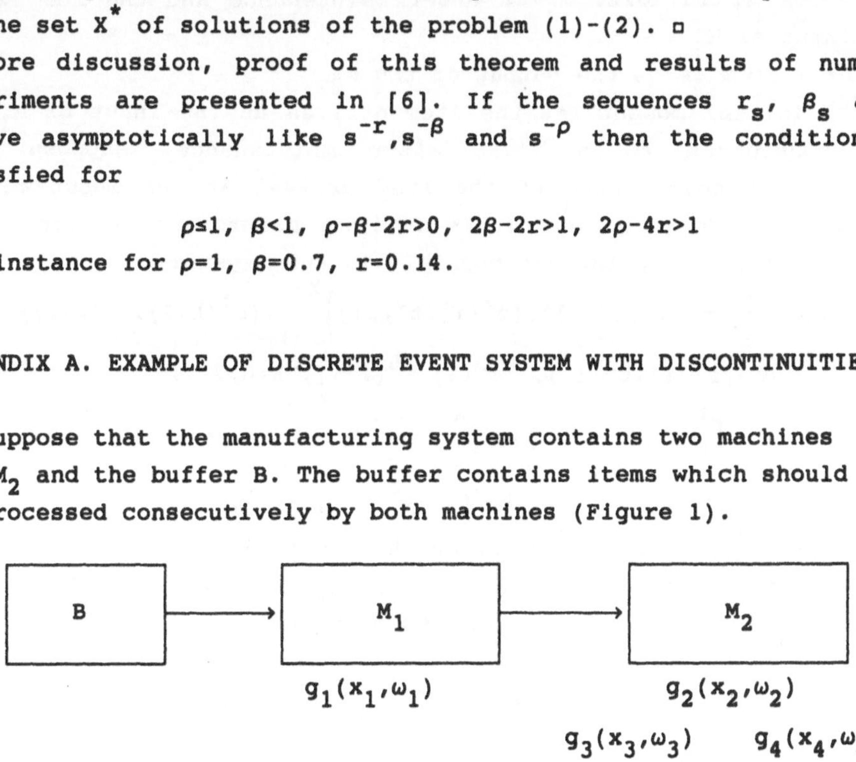

$$g_1(x_1,\omega_1) \qquad\qquad g_2(x_2,\omega_2)$$
$$g_3(x_3,\omega_3) \qquad g_4(x_4,\omega_4)$$

Figure 1.

The processing time of each machine is $g_i(x_i,\omega_i)$, $i = 1,2$, $x_i \in \mathbb{R}^1$, ω_i is distributed uniformly on $[0,1]$. If, for example, processing time is distributed exponentially and x_i is processing rate then

$$g_i(x_i,\omega_i)=-\frac{1}{x_i}\ln(1-\omega_i)$$

The performance capability of the second machine can deteriorate and is monitored by separate process. If it is detected that the second machine has deteriorated below certain level and the machine is idle then maintenance is started. If it is busy then maintenance is started immediately after finishing the job. If an item arrives at the input of the second machine during maintenance period then it waits till the end of maintenance and immediately after that the processing is started. Time elapsed between the end of one maintenance period and detection of necessity for another maintenance is $g_3(x_3,\omega_3)$, the length of maintenance is $g_4(x_4,\omega_4)$. Suppose for simplicity that the buffer contains only one item. Then the system can be in the following states:

z^1 — M_1 is busy, M_2 is idle and ready for a job

z^2 — M_1 is busy, M_2 is under maintenance

z^3 — M_1 is idle, M_2 is busy

z^4 — M_1 is idle, M_2 is under maintenance and the item waits at the input of M_2

z^5 — item is at the output of the M_2.

At the initial moment t=0 the item arrives at the input of M_1 and M_2 is considered to be just after maintenance. Suppose that probability of coincidence of the item arrival at the input of the second machine and detection of the need for maintenance is zero. Then the following sample paths are possible in this system:

$$U^{1k}(x,\omega)=\left\{\left[(z^1(i),t^1(i)),(z^2(i),t^2(i))\right]_{i=1}^k,(z^1(k+1),t^1(k+1)),\right.$$
$$\left.(z^3(1),t^3(1)),\ (z^5(1),t^5(1))\right\},\ k=0,2,\ldots$$

$$U^{2k}(x,\omega)=\left\{\left[(z^1(i),t^1(i)),(z^2(i),t^2(i))\right]_{i=1}^k,(z^4(1),t^4(1)),\right.$$
$$\left.(z^3(1),t^3(1)),\ (z^5(1),t^5(1))\right\},\ k=1,2,\ldots$$

where $(z^j(i),t^j(i))$ denotes event which consists of the i-th transition to the state j from the beginning of simulation, in order to simplify notations we omitted dependence on (x,ω). Here

$$t^1(1)=0,\ t^1(k)=G(k-1,x,\omega),\ k\geq2,\ G(k,x,\omega)=\sum_{i=1}^k\left(g_3(x_3,\omega_3^i)+g_4(x_4,\omega_4^i)\right)$$

$$t^2(k)=t^1(k)+g_3(x_3,\omega_3^k),\ t^3(1)=\begin{cases}g_1(x_1,\omega_1^1)&\text{for path }z^{1k}(x,\omega)\\G(k,x,\omega)&\text{for path }z^{2k}(x,\omega)\end{cases}$$

$$t^4(1)=g_1(x_1,\omega_1^1),\ t^5(1)=t^3(1)+g_2(x_2,\omega_2^1),\qquad\qquad\text{(A.1)}$$

The path $U^{1k}(x,\omega)$ is taken if $(x,\omega) \in \theta_{1k}$ and the path $U^{2k}(x,\omega)$ is taken in the case $(x,\omega) \in \theta_{2k}$, where

$$\theta_{1k} = \left\{ (x,\omega) : \ G(k,x,\omega) \le g_1(x_1,\omega_1^1) \le G(k,x,\omega) + g_3(x_3,\omega_3^{k+1}) \right\}, \tag{A.2}$$

$$\theta_{2k} = \left\{ (x,\omega) : \ G(k-1,x,\omega) + g_3(x_3,\omega_3^k) < g_1(x_1,\omega_1^1) < G(k,x,\omega) \right\}, \tag{A.3}$$

Suppose that the objective function is the weighted sum of thruput and cost terms. The thruput term in this case is the time of arrival for the first time at the state z^5, since only one item is in the buffer, i.e. it equals $t^5(1)$. Summarizing (A.1)- (A.3) we obtain:

$$F(x) = F^1(x) + F^2(x), \quad F^1(x) = E_\omega f(x,\omega),$$

$$f(x,\omega) = \begin{cases} g_1(x_1,\omega_1^1) + g_2(x_2,\omega_2^1) & \text{if} \quad (x,\omega) \in \theta_{1k} \\ G(k+1,x,\omega) + g_2(x_2,\omega_2^1) & \text{if} \quad (x,\omega) \in \theta_{2(k+1)} \end{cases} \tag{A.4}$$

where $k=0,1,\ldots$ and $G(0,x,\omega)=0$. Therefore the function $f(x,\omega)$ is discontinuous with respect to (x,ω), but it is differentiable on each set θ_{1k}, θ_{2k} if $g_i(x_1,\omega_i)$ are differentiable. Note that the function $F(x)$ may be differentiable too, depending on the properties of $g_i(x_1,\omega_i)$. In particular, it would be differentiable in the case when $g_i(x_1,\omega_i)$ are distributed exponentially.

Thus, even in such simple example as this, there are infinite number of sets in continuity partition defined by (A.2)-(A.3). If we apply differentiation scheme to obtain estimate of the gradient of $F^1(x)$ we should take into account discontinuities on the boundaries of these sets. It is not sufficient to take the sample derivative of the function under the integral sign. In order to see this let us compute partial derivative of $F_1(x)$ with respect to x_1. Let us denote

$$a(k,x,\omega) = a(k) : \ g_1(x_1,a(k,x,\omega)) = G(k,x,\omega) + g_3(x_3,\omega_3^{k+1}), \quad k \ge 0$$

$$b(k,x,\omega) = b(k) : \ g_1(x_1,b(k,x,\omega)) = G(k,x,\omega), \quad k \ge 1, \ b(0,x) = 0$$

Then

$$F_{x_1}^1(x) = \sum_{k=0}^{\infty} \int_0^1 \int_{b(k)}^{a(k)} g_{1x_1}(x_1,\omega_1^1) d\omega_1^1 d\omega_3^1 \cdots d\omega_3^{k+1} d\omega_4^1 \cdots d\omega_4^k +$$

$$\sum_{k=0}^{\infty} \int_0^1 (a_{x_1}(k)(g_1(x_1,a(k)) - G(k+1,x,\omega)) d\omega_3^1 \cdots d\omega_3^{k+1} d\omega_4^1 \cdots d\omega_4^{k+1} \tag{A.5}$$

Now let us try to compute the sample derivative using only one sample path, which amounts to differentiation of $f(x,\omega)$. We obtain

$$f_{x_1}(x,\omega) = \begin{cases} g_{1x_1}(x_1,\omega_1^1) & \text{if} \quad (x,\omega) \in \theta_{1k} \\ 0 & \text{if} \quad (x,\omega) \in \theta_{2(k+1)} \end{cases} \tag{A.6}$$

Note, that under general assumptions this derivative exists almost everywhere. Taking expectation in (A.6) we would obtain only the first term in (A.5) and would lose the second term, which appears due to discontinuities.

APPENDIX B. EXAMPLE FOR THE SCORE FUNCTION METHOD

Let us consider the queuing network on the Figure 2:

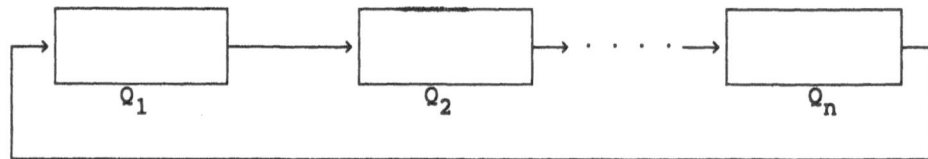

Figure 2.

which consists of n servers $Q_1, \ldots, Q_n$ connected consecutively and several customers which circulate in the system. For the sake of simplicity let us consider the case when there is only one customer in the system, which enters the server Q_1, passes consecutively through all servers and returns to the server Q_1. The service time t_i at each server is distributed exponentially:

$$P(t_i \geq T) = 1 - e^{-Tx}$$

where x is the service rate, and t_i, i=1:n are independent. Let us estimate the derivative of average time needed for a customer which starts from the server Q_1 to pass through the system and return to Q_1 with respect to the service rate x. Note, that x is the same for all servers.

The average time is:

$$\bar{t}(x) = \int \cdots \int \sum_{i=1}^{n} t_i \prod_{i=1}^{n} \left(x e^{-x t_i} \right) dt_1 \cdots dt_n$$

The derivative of average time:

$$\bar{t}_x(x) = \int \cdots \int \sum_{i=1}^{n} t_i \sum_{j=1}^{n} \frac{1 - x t_j}{x} \prod_{i=1}^{n} \left(x e^{-x t_i} \right) dt_1 \cdots dt_n$$

The likelihood ratio/score function estimate of the derivative:

$$t_x^{SF} = \sum_{i=1}^{n} t_i \sum_{j=1}^{n} \frac{1 - x t_j}{x}$$

Its expectation:

$$\mathbb{E} t_x^{SF} = -\frac{n}{x^2} = \bar{t}_x(x)$$

Its variance:

$$R^{SF} = \mathbb{E}(t_x^{SF} - \mathbb{E} t_x^{SF})^2 = \mathbb{E}(t_x^{SF} - \bar{t}_x(x))^2 = \frac{1}{x^4}(n^3 + 6n^2 + 6n)$$

Let us consider now the finite difference estimate. We would use the simplest forward finite difference scheme. In this case

$$t_x^{FD} = \frac{\sum\limits_{i=1}^{n} t_i(x+\delta) - \sum\limits_{i=1}^{n} t_i(x)}{\delta} = \frac{1}{\delta} \sum_{i=1}^{n} (t_i(x+\delta) - t_i(x))$$

where by $t_i(x)$ we denoted the observation of the processing time made with the value of processing rate equal to x.

The expectation of this estimate:

$$\mathbb{E} t_x^{FD} = \frac{n}{\delta} \left(\frac{1}{x+\delta} - \frac{1}{x} \right) = - \frac{n}{x(x+\delta)}$$

Unlike t_x^{SF}, this estimate is biased. As the measure of accuracy of this estimate let us consider $R^{FD} = \mathbb{E}(t_x^{FD} - \bar{t}_x(x))^2$ which would take into account both systematic and random error:

$$R^{FD} = \mathbb{E}(t_x^{FD} - \bar{t}_x(x))^2 = \frac{n}{\delta^2} \left(\frac{1}{(x+\delta)^2} + \frac{1}{x^2} \right) + \frac{n^2\delta^2}{x^4(x+\delta)^2}$$

Here are the values of R^{FD} and R^{SF} for some fixed values of n,x,δ

$$\delta=0.1, \quad n=4, \quad x=0.5, \quad R^{FD}=2718.2, \quad R^{SF}= 2944$$

$$\delta=0.2, \quad n=1, \quad x=0.5, \quad R^{FD}= 152.3, \quad R^{SF}= 208$$

$$\delta=0.3, \quad n=2, \quad x=1, \quad R^{FD}= 35.6, \quad R^{SF}= 44$$

I.e. in this example even for small values of n and reasonable values of δ the finite differences estimate could be more accurate then the likelihood ratio/score function estimate. The problems of the score function estimate would be even more pronounced if there is more then one customer in the system, since in this case it would be more difficult to define a regenerative period. In the absence of regenerative property it would be necessary to increase the number of simulation steps which would increase the variance of the score function estimate still further.

REFERENCES.

1. F.Archetti, A.Gaivoronski and A.Sciomachen, Sensitivity analysis and optimization of stochastic Petri Nets, Preprint, University of Milano, 1990.

2. V.M.Aleksandrov, V.I.Sysoyev and V.V.Shemeneva, Stochastic optimization, Eng.Cybern., v.5, 1968, p.11-16.

3. X.R.Cao, Convergence of parameter sensitivity estimates in a stochastic experiment, IEEE Transactions on Automatic Control, v. AC-30, No.9, 1985, p.845-853.

4. M.A.Crane and D.L.Iglehart. Simulating stable stochastic systems. III. Regenerative processes and discrete-event simulations,

Oper. Res. vol.23, 1975, p.33-45.

5. Yu.Ermoliev, Methods of Stochastic Programming, Nauka, Moscow, 1976 (in Russian).

6. Yu. Ermoliev and A.A.Gaivoronski, Stochastic programming techniques for optimization of discrete event systems, Annals of Operations Research, 1991.

7. Yu.Ermoliev, Optimization of discrete event systems described by semi-Markov processes, Lecture at the conference "Computationally intensive methods in Simulation and Optimization", Vienna, August 1990.

8. Yu.Ermoliev and Wets, R.J.-B. eds. Numerical Techniques for Stochastic Optimization, Springer-Verlag, Berlin, 1988.

9. A.A.Gaivoronski, Approximation methods of solution of stochastic programming problems,- Kibernetika, 2, 1982 (in Russian, English translation in: Cybernetics, v.18, No.2).

10. A.A.Gaivoronski, Interactive Program SQG-PC for Solving Stochastic Programming Problems on IBM PC/XT/AT Compatibles. User Guide. Working Paper WP-88-11, IIASA, Laxenburg, 1988.

11. A.A.Gaivoronski, Augmented Perturbation Analysis for optimization of discrete event systems, Preprint, Institute of Cybernetics, Kiev, 1990.

12. P.Glasserman and W.B.Gong. Smoothed perturbation analysis for a class of discreet-event systems. IEEE Trans. on Automatic Control, 35(11): 1218-1230, 1990.

13. P.W.Glynn, Optimization of stochastic systems, in: Proceedings of 1986 Winter Simulation Conference, 1986.

14. P.W.Glynn and J.L.Sanders, Monte Carlo Optimization of Stochastic Systems: Two New Approaches. Proc. 1986 ASME Computing in Engineering Conference, (Chicago, IL) 1986.

15. W.B.Gong and Y.C.Ho, Smoothed (Conditional) Perturbation Analysis of discrete event dynamic systems, IEEE Transactions on Automatic Control, AC-32, 1987, 856-866.

16. W.B.Gong, C.G.Cassandras and J.Pan, Perturbation analysis of a multiclass queueing system with admission control. IEEE Transactions on Automatic Control, vol.36, No.6, 1991, p.707-723.

17. P.Heidelberger, Xi-Ren Cao, M.A.Zazanis and R.Suri. Convergence properties of Infinitesimal Perturbation Analysis estimates, Management Science, v.34, No.11, 1988.

18. C.A.R.Hoare, Communicating Sequential Processes. Englewood Cliffs, NJ: Prentice-Hall International, 1985.

19. Y.C.Ho, Performance evaluation and perturbation analysis of discrete event dynamic systems. IEEE Transactions on Automatic

Control, vol. AC-32, No.7, 1987, p.563-572.

20. Y.C.Ho and S.Li, Extensions of Infinitesimal Perturbation Analysis, IEEE Transactions on Automatic Control, AC-33, 1988, p.427-438.

21. Y.C.Ho, M.A.Eyler and T.T.Chien. A gradient technique for general buffer storage design in a serial production line. Int. J. Prod. Res., v.17, No.6, 1979, p.557-580.

22. Y.C.Ho, (ed.). A selected and annotated bibliography on perturbation analysis, Lecture Notes in Control and Information Sciences, Vol.103, Springer-Verlag, 1987, pp.162-178.

23. Y.C.Ho, L.Shi, L. Dai and W.Gong. Optimizing discrete event dynamic systems via the gradient surface method, Manuscript, Harward University, 1990.

24. J.E.Hopcroft and J.D.Ullman, Introduction to Automata Theory, Languages and Computation, Reading, MA: Addison-Wesley, 1979.

25. J.Q.Hu and Y.-C.Ho. An infinitesimal perturbation analysis algorithm for a multiclass G/G/1 queue. OR Letters, 9:35-44, 1990.

26. K.Inan and P.Varaiya, "Finitely recursive process models for discrete event systems", IEEE Transactions on Automatic Control, v.33, No.7, 1988, p. 626-639.

27. S.H.Jacobson and Schruben L.W., Techniques for simulation response optimization, Operations Research Letters, Feb. 1989, 1-9.

28. P.Kall, Stochastic Linear Programming, Springer Verlag, Berlin, 1976.

29. J.Kiefer and J.Wolfowitz. Stochastic estimation of a maximum of a regression function, Ann. Math. Statist. 23, 1952, 462-466.

30. H.Kushner, and D.S.Clark. Stochastic Approximation for Constrained and Unconstrained Systems, Appl. Math. 26, Springer, 1978.

31. K.Marti, Lecture at the Conference "Computationally intensive methods in Simulation and Optimization", IIASA, Laxenburg, 1990.

32. D.L.McLeich and S.Rollans. Conditioning for variance reduction in estimating the sensitivity of simulations. To be published in Annals of Operations Research, 1992.

33. M.S.Meketon, "Optimization in simulation: A survey of recent results", Proceedings of the 1987 Winter Simulation Conference, A.Thesen, H.Grant, W.David Kelton (eds), pp. 58-61.

34. R.H.Myers, Response Surface Methodology, Allyn and Bacon, Boston, Massachusetts, 1987.

35. J.L.Peterson, Petri Net Theory and the Modelling of Systems. Englewood Cliffs, NJ: Prentice Hall, 1981.

36. G.Ch.Pflug, On line optimization of simulated Markovian processes. Mathematics of OR, 1990.

37. G.Ch.Pflug, Derivatives of probability measures - concepts and applications to the optimization of stochastic systems, in: Discrete Events Systems: Models and Applications. IIASA Conference, Sopron, Hungary, August 3-7, 1987, P.Varaiya and A.B.Kurzhanski (eds.), Lecture Notes in Control and Information Sciences, Springer Verlag, 1988, p.162-178.

38. G.Ch.Pflug, Optimization of simulated discrete event processes, Preprint TR-ISI/Stamcom 87, University of Vienna, 1990.

39. M.I.Reiman and A.Weiss, Sensitivity analysis via likelihood ratios, in: Proceedings of the 1986 Winter Simulation Conference, 1986, p.285-289.

40. R.Y.Rubinstein, The score function approach of sensitivity analysis of computer simulation models. Math. and Computation in Simulation, vol.28, 1986, p.351-379.

41. R.Y.Rubinstein, Monte Carlo Optimization, Simulation and Sensitivity Analysis of Queuing Networks. New York, Wiley, 1986.

42. R.Y.Rubinstein. How to optimize discrete-event systems from a single sample path by the score function method. Annals of Operations Research 27 (1991) 175-212.

43. R.Suri, Infinitesimal Perturbation Analysis of General Discrete Event Systems, J. Assoc. Comput. Mach., 34, 1987, 686-717

44. R.Suri, Perturbation Analysis: The State of the Art and Research Issues Explained via the GI/G/1 Queue, Proceedings of the IEEE, v.77, No.1, 1989, 114-137.

45. R.Suri and Y.T.Leung, Single run optimization of a SIMAN model for automatic assembly systems, Proceedings of the 1987 Winter Simulation Conference, A.Thesen, H.Grant, W.D.Kelton (eds.)

46. R.J.-B.Wets, Stochastic programming: solution techniques and approximation schemes, in: Mathematical Programming: the State of the Art, 1982, eds. A.Bachem, M.Grotschel and B.Korte, Springer Verlag, 1983, p.566-603.

47. W.Whitt, Continuity of generalized semi-Markov process, Math. Oper. Research, vol.5, 1980, p.494-501.

SENSITIVITY ANALYSIS OF SIMULATION EXPERIMENTS: REGRESSION ANALYSIS AND
STATISTICAL DESIGN

Jack P.C. Kleijnen
Katholieke Universiteit Brabant
(Tilburg University)
Tilburg, Netherlands

ABSTRACT

This tutorial gives a survey of strategic issues in the statistical design and analysis of experiments with deterministic and random simulation models. These issues concern what-if analysis, optimization, and so
on. The analysis uses regression (meta)models and Least Squares. The design uses classical experimental designs such as 2^{k-p} factorials, which
are efficient and effective. If there are very many inputs, then special
techniques such as group screening and sequential bifurcation are useful.
Applications are discussed.

INTRODUCTION

Simulation is a mathematical technique that is applied in all
scientific disciplines that use mathematical modeling, ranging from socio-
logy to astronomy; also see Karplus [1]. It is a very popular technique
because of its flexibility, simplicity, and realism. By definition, simulation involves experimentation, namely with the model of a real system.
Consequently it requires on appropriate design and analysis. For real
systems mathematical statistics has been applied since the 1930's: Sir
Ronald Fisher focussed on agricultural experiments in the 1930's; George
Box concentrated on chemical experimentation, since the 1950's; see [2].
Tom Naylor organized a conference on the design of simulation experiments
back in 1968; see [3]. In 1974/75 my first book [4] covered both the 'tac-
tical' and 'strategic' issues of experiments with random and deterministic

simulation models. The term _tactical_ was introduced into simulation by Conway [5]; it refers to issues such as runlength and variance reduction, which arise only in random simulations such as queuing simulations. _Strategic_ questions are: which combinations of input variables should be simulated, and how can the resulting output be analyzed? Obviously strategic issues arise in both random and deterministic simulations. Mathematical statatistics can be applied to solve these questions, also in deterministic simulation; see the recent publications [6]. [7], and [8]. This contribution focusses on these strategic issues in simulation experiments.

Strategic issues concern problems that are also addressed under names like model _validation_, _what-if_ analysis, _goal_ seeking, and _optimization_; see table 1, reproduced from my recent book [6, p. 136]. We shall return to this table.

REGRESSION METAMODELS

Before the systems analyst starts experimenting with the simulation model, he (or she) has accumulated _prior_ knowledge about the system to be simulated: he may have observed the real system, tried different models, debugged the final simulation model, and so on. This tentative knowledge isformalized in a regression or Analysis of Variance (ANOVA) model. ANOVA models are elementary in the statistical theory of experimental design: Sums of Squares (SS's) are compared through the F test to detect significant main effects and interactions; see below. The simplest ANOVA models can be easily translated into regression models; see [6, pp. 263-293]. Because regression analysis is more familiar than ANOVA is, we shall use regression terminology henceforth.

Table 1: Terminology

Computer program	Simulation model	Regression model	User view
Output	Response	Dependent variable y	Result
Input	Parameter	Independent variable x	Environment
	Variable		
	Enumeration	Continuous	Validation
			Risk Analysis
	Function	Discrete	Controllable
	Scenario	Binary	Optimization
			Goal output (control)
			Satisfy (what-if)
	Behavioral re- lationship		

So prior knowledge is formalized in a <u>tentative regression model</u>. In other words, this model is tested later on to check its validity as we shall see. The regression model specifies which <u>inputs</u> seem important, which <u>interactions</u> among these inputs seem important, and which <u>scaling</u> seems appropriate. We shall discuss these items next.

Table 1 showed that 'inputs' are not only parameters and variables but may also be 'behavioral relationships'; that is, a module of the simulation model may be replaced by a different module. In the regression model such a qualitative change is represented by one or more binary (0,1) variables. Note that 'inputs' are called 'factors' in experimental design terminology. 'Interaction' means that the effect of a factor depends on the values (or 'levels') of another factor:

$$y = \beta_0 + \sum_{j=1}^{k} \beta_j x_j + \sum_{j=1}^{k} \sum_{g=1}^{k} \beta_{jg} x_j x_g +$$

$$+ \sum_{j=1}^{k} \sum_{g=1}^{k} \sum_{h=1}^{k} \beta_{jgh} x_j x_g x_h + \ldots + e, \tag{1}$$

where y is the simulation response; β_0 is the overall or grand mean; β_j is the main or first-order effect of factor j (j = 1,..., k); β_{jg} is the two-factor interaction between the factors j and g (g = 1,..., k; g ≠ j); β_{jj} is the quadratic effect of factor j; β_{jgh} is the three-factor interaction among the factors j, g, and h (h = 1,..., k; h ≠ g ≠ j); and so on; e denotes 'fitting 'errors' or noise. Under certain strict mathematical conditions the 'response curve' in Eq. (1) is a Taylor series expansion of the simulation model $y(x_1,...,x_k)$. Unfortunately these conditions do not hold in simulation. Therefore we propose to start with an initial model that excludes interactions among three or more factors: such high-order interactions are popular in ANOVA but they are hard to interpret. The purpose of the regression model is to guide the design of the simulation experiment and to interpret the resulting simulation data; a regression model without high-order interactions suffices, as we observed repeatedly in practice.

The regression variables x in Eq. (1) may be <u>transformations</u> of the original simulation parameters and variables; for example, $x_1 = \log(z_1)$ where z_1 denotes the original simulation input. <u>Scaling</u> is also important: if the lowest value of z_1 corresponds with $x_1 = -1$ and its highest value corresponds with $x_1 = +1$, then β_1 measures the relative importance of factor 1 when that factor ranges over the experimental area. In optimization we explore the response curve only locally if we use <u>Response Surface Methodology</u> (RSM). Then the local regression model is a first-order model:

$$y = \gamma_0 + \sum_j \gamma_j z_j + e, \tag{2}$$

where the importance of factor j at $\bar{z}_j$, the midpoint of the local experiment, is measured by $\gamma_j \bar{z}_j$; obviously $\bar{z}_j = \Sigma_{j-1}^n z_{ij}/n = (L_j + H_j)/2$ where $L_j \leq z_{ij} \leq H_j$ with local experimental area $[L_1, H_1] \times \cdots \times [L_k, H_k]$; z_{ij} denotes the value of factor j in simulation run or observation i. See [9] and [2].

In any experiment the analyst uses a model such as Eq. (1), explicitly or implicitly. For example, if he changes one factor at a time, then (implicitly) he assumes all interactions $(\beta_{jg}, \beta_{jgh}, \ldots)$ to be zero. Of course it is better to make the regression model explicit and to find a design that fits that model, as we shall see next. But first note that we call the regression model a metamodel because it models the input/output behavior of the underlying simulation model.

EXPERIMENTAL DESIGN

Based on a tentative regression (meta)model we select an experimental design. The design matrix $D = (d_{ij})$ specifies the n combinations of the k factors that are to be simulated. (In multi-stage experimentation such as RSM that set of n combinations is followed by a next set.) Classical statistical theory gives designs that are 'efficient' and 'effective'. Efficiency means that the number of combinations or simulation runs is 'small'. Suppose there are Q effects in the regression model. The number of runs should satisfy the condition $n \geq Q$; for example, we need $k + 1$ runs if there are no interactions at all. So we may do one base run (say) $x'_0 = (-1, -1, \ldots, -1)$; and then we change one factor at a time: $x'_1 = (+1, -1, \ldots, -1)$, $x'_2 = (-1, +1, -1, \ldots, -1), \ldots, x_k = (-1, \ldots, -1, +1)$; see table 2 for $k = 3$. However, to estimate the effects $\beta' = (\beta_0, \beta_1, \ldots, \beta_k)$ we fit a curve to the simulation data (X, y) where $X = (1, D)$ in the first-order model; 1 denotes a vector of n ones. The classical fitting criterion is Least Squares. This criterion yields the estimator

$$\hat{\beta} = (X' X)^{-1} X' y. \tag{3}$$

Now consider the classical 2^{3-1} design of Table 2. The corresponding X is orthogonal, so (3) reduces to the scalar expression

$$\hat{\beta}_{j'} = \Sigma_{i=1}^{n} x_{ij'} y_i/n \qquad (j' = 0,1,\ldots,k).$$ (4)

Table 2. Two designs for three factors.

Run	One at a time			2^{3-1} Design		
	d 1	d 2	d 3	d 1	d 2	d 3
1	-	-	-	-	-	+
2	+	-	-	+	-	-
3	-	+	-	-	+	-
4	-	-	+	+	+	+

How can we choose between the two designs of table 2? Classical theory assumes that the fitting errors e are white noise: e is normally, and independently distributed with zero mean and constant variance (say) σ^2. Then (3) yields the variance-covariance matrix

$$cov(\hat{\underline{\beta}}) = (X' X)^{-1} \sigma^2 .$$ (5)

An orthogonal matrix X is optimal: it minimizes var $(\hat{\beta}_j)$, the elements on the main diagonal of Eq. (5); see [6, p.335]. There are straightforward procedures for deriving design matrices, if $n = 2^{k-p}$ with $(p=0,1,\ldots)$; for other n values results are tabulated; see [2] and [6].

So the classical designs are efficient under the white noise assumption (recent research uses alternative assumptions; see Sachs et al. [7]). Moreover, these designs are effective: they permit the estimation of interactions. If we allow for two-factor interactions, then the number of effects becomes $Q = 1 + k + k(k-1)/2$. If k is small, we may simulate $n \geq Q$ combinations; for example, if $k = 5$ then a 2^{5-1} design is suitable. (If k is large, then we may hope that some factors will turn out to give nonsignificant main effects; we may assume that factors without main effects have no two-factor interactions either; there are designs that yield unbiased estimators for main effects with $n = 2 \, k < 1 + k + k(k-1)/2$; see [6, pp. 303 - 309], [9].) If the factors are quantitative, then a second-order regression model has k quadratic effects too. Then n must increase, and more than two levels per factor must be simulated: RSM designs; see [2] and [6].

SCREENING

For didactic reasons we discuss 'screening' designs <u>after</u> classical experimental designs. In practice the simulation model has a great many factors that may be important; of course the analyst assumes that only a few factors are really important: parsimony. So in the beginning of a simulation study it is necessary to search for the few really important factors among the many conceivably important factors: <u>screening</u>.

Classical textbooks do not discuss screening situations, because in real-life experiments it is impossible to control (say) a hundred factors. In simulation, however, we perfectly control all inputs and we indeed use models with many inputs. One approach is <u>group screening</u>, introduced in the early 1960's by Watson, Jacoby and Harrison, Li, and Patel. This technique aggregates the many individual factors into a few group factors. Some simulation applications can be found in [6, p.327]. Recently Bettonvil [10] further developed group screening into <u>sequential bifurcation</u>, a very efficient technique that accounts for white noise and interactions. He applied this technique to an ecological model with nearly 300 factors.

REGRESSION ANALYSIS: TECHNICALITIES

Eq. (3) gave the Ordinary Least Squares (OLS) estimator $\hat{\beta}$. In deterministic simulation that estimator may suffice, although Sachs et al. [7] give a better estimator if the white noise assumption is dropped (and replaced by a stationary covariance assumption). In random simulation the classical assumptions seldom hold. If the response variances differ with the inputs (as the response means do), then Weighted Least Squares (WLS) is better. If common random numbers drive the various factor combinations, then Generalized Least Squares (GLS) is best. See [6, pp. 161-175].

Once the regression model is calibrated (that is, the parameters β are estimated), the metamodel's validity must be tested. For deterministic simulation models we propose cross validation: delete factor combination i (x_i', y_i); reestimate β from the ramaining simulation data (X_{-i}, y_{-i}); predict the deleted response y_i through the reestimated regression model $(\hat{y}_i = \hat{\beta}'_{-i} x_i)$; "eyeball" the relative prediction errors $\hat{y}_i/y_i$: are these errors acceptable to the user?

In random simulation we prefer Rao's adjusted lack-of-fit F-test: the estimated response variances and covariances are compared with the residuals $(\hat{y} - y)$; see [11].

SOME APPLICATIONS

Applications of our approach are getting numerous. For example, a simple - but realistic - case study concerns a Flexible Manufacturing System (FMS). Input to the FMS simulation is the 'machine mix', that is, the number of machines of type i with $i = 1,...,4$. Intuitively selected combinations of these four inputs give inferior results when compared with a classical design. The throughput predicted by the simulation is analyzed through two different regression models. These models are validated. A regression model in only two inputs but including their interaction, gives valid predictions and sound explanations [12].

Another application concerns a decision support system (DSS) for production planning, developed for a Dutch company. To evaluate this DSS, a simulation model is built. The DSS has 15 control variables that are to be optimized. The effects of these 15 variables are investigated, using a sequence of classical designs. Originally, 34 response variables were distinguished. These 34 variables, however, can be reduced to one criterion variable, namely productive machine hours, that is to be maximized, and one commercial variable measuring lead times, that must satisfy a certain side-condition. For this optimization problem the Steepest Ascent technique is applied to the experimental design outcomes. See [13].

A final case study concerns a set of deterministic ecological simulation models that require sensitivity analysis to support the Dutch government's decision making. First results for a model of the 'greenhouse' effect are given in [14]; additional results are given in [10].

CONCLUSIONS

Experimental design and regression analysis are statistical techniques that have been widely applied in the design and analysis of data obtained by real life experimentation and observation. In simulation, these techniques are gaining popularity: a number of case studies have been published. The techniques need certain adaptations to account for the peculiarities of deterministic and random simulations.

REFERENCES

[1] Karplus, W.J., The spectrum of mathematical models. Perspectives in Computing, 3, no. 2, May 1983, pp. 4-13.

[2] Box, G.E.P. and N.R. Draper, Empirical Model-Building And Response Surfaces. John Wiley & Sons, New York, 1987.

[3] Naylor, T.H., editor, The Design of Computer Simulation Experiments, Duke University Press, Durham, N.C. 1969.

[4] Kleijnen, J.P.C., Statistical Techniques in Simulation. Marcel Dekker, Inc., New York, 1974/1975. (Russian translation, Moscow, 1978.)

[5] Conway, R.W., Some tactical problems in digital simulation. <u>Manage-ment Science</u>, <u>10</u>, no. 1, Oct. 1963, p. 47-61.

[6] Kleijnen, J.P.C., <u>Statistical Tools for Simulation Practitioners</u>, Marcel Dekker, Inc., New York, 1987.

[7] Sachs, J., W.J. Welch, T.J. Mitchell, and H.P. Wynn, Design and ana-lysis of computer experiments, <u>Statistical Science</u>, <u>4</u>, no. 4, 1989, pp. 409-435.

[8] Kleijnen, J.P.C., <u>Statistics and deterministic simulation: Why not?</u> Katholieke Universiteit Brabant (Tilburg University), Tilburg, 1990.

[9] Bettonvil, B. and J.P.C. Kleijnen, Measurement scales and resolution IV designs. <u>American Journal of Mathematical and Management Sciences</u> (accepted).

[10] Bettonvil, B., <u>Detection of Important Factors by Sequential Bifurca-tion</u>. Doctoral dissertation, Katholieke Universiteit Brabant (Tilburg University). Tilburg, November 1990.

[11] Kleijnen, J.P.C., Regression metamodels for simulation with common random numbers: comparison of techniques. Katholieke Universiteit Brabant (Tilburg University), Tilburg, January 1990.

[12] Kleijnen, J.P.C. and C.R. Standridge, Experimental design and regres-sion analysis in simulation: an FMS case study. <u>European Journal of Operational Research</u>, <u>33</u>, 1988, pp. 257-261.

[13] Kleijnen, J.P.C., <u>Simulation and Optimization in Production Planning: a Case Study</u>, Katholieke Universiteit Brabant (Tilburg University), Tilburg, 1988.

[14] Rotmans, J. and O.J. Vrieze, Metamodelling, a sensitivity analysis on the greenhouse effect. <u>European Journal of Operational Research</u> (ac-cepted).

A STOCHASTIC OPTIMIZATION APPROACH FOR
TRAINING THE PARAMETERS IN NEURAL NETWORKS

Norio Baba

Information Science, Osaka Educational
Univ., Jonan 3-1-1, Ikeda City, 563,
Japan.

Abstract   Recently, back-propagation method has often been applied
to adapt artificial neural networks for various pattern classification
problems.   However, an important limitation of this method is that it
sometimes fails to find a global minimum of the total error function of
neural network.   In this paper, a hybrid algorithm which combines the
modified back-propagation method and the random optimization method is
proposed in order to find the global minimum of the total error function
of neural network in a small number of steps.   It is shown that this
hybrid algorithm ensures convergence to a global minimum with probability
1 in a compact region of weight vector space.   Further, several computer
simulation results dealing with the problem of forcasting air pollution
density, stock price, and etc. are given.

I. INTRODUCTION

In recent years, neural network computing has been studied quite extensively by
many researchers and various fruitful results have been obtained.   In particular,
the back-propagation method (BP method) proposed by Rumelhart and McClelland [13]is
one of the most stimulating products and has given great impact to the development
of this area.

However, the most important limitations of this method are:

(1)   It does not necessarily ensure convergence to the global minimum of the total
error function of neural network.

(2)   Since it consists of a steepest descent method [9] which uses fixed step size
rule in the line search of the objective function, it does not necessarily
ensure monotonous decreasing property of the total error function, and
therefore, it does not ensure convergence even to a local minimum of the total
error function of neural network.

It has been suggested [4],[8] that some stochastic optimization approach should be
introduced to solve the first problem.   To solve the second problem, we need to
improve the original BP method by performing a line search in each step of calculation.

In this paper, a hybrid algorithm which combines the modified BP method and the random optimization method of Solis & Wets [12] is proposed in order to find the global minimum of the total error function of neural network in a small number of steps.

Section 1 of this paper is devoted to an explanation of the original BP method Several problems inherent in this method are also pointed out. In Section 2 of this paper, a new modified back-propagation method which performs minimization using the conjugate gradient method [6] , [9] and carries out a simple line search is proposed. Further, this method is combined with the random optimization method of Solis & Wets [12] in order to construct a hybrid algorithm for finding the global minimum of the total error function of neural network in a small number of steps. A theorem which shows that our proposed hybrid algorithm ensures convergence with probability 1 to the global minimum of the total error function of neural network is also given. In the final part of this paper, several computer simulations dealing with the problem of forecasting air pollution density, prices of stocks of several enterprises, and etc. are given.

II. LEARNING OF THE WEIGHTS OF NEURAL NETWORK BY THE BACK-PROPAGATION METHOD

The back-propagation method proposed by Rumelhart and McClelland [13] has often been applied to adapt artificial neural networks for various actual pattern classification problems. In this section, this method is briefly introduced.

<u>BP Method</u>   Let us consider the following input-output relations (1) and (2):

$$x_{ip}(w) = f_i(z_{ip}(w)) \tag{1}$$

$$z_{ip}(w) = \sum_j w_{ij} y_{jp}(w) \tag{2}$$

, where w, $y_{jp}(w)$, and $x_{ip}(w)$ denote the weight vector of neural network, the jth output from the (s-1)th layer corresponding to the pth input pattern, and the output of the ith neuron of the sth layer corresponding to the pth input pattern, respectively.

The error function $E_p(w)$ corresponding to the pth input pattern and the total error function $E(w)$ describing the total difference between ideal output and real output are defined by (3) and (4):

$$E_p(w) = \frac{1}{2} \sum_i (\bar{x}_{ip} - d_{ip})^2 \tag{3}$$

$$E(w) = \sum_p E_p(w) \tag{4}$$

where $\bar{x}_{ip}(w)$ and d_{ip} are the actual output from the ith unit of the output layer corresponding to the pth pattern and the signal (ideal output) shown by the teacher, respectively.

The original BP method updates the weight vector w to the steepest descent direction of $E_p(w)$. However, strictly speaking, this method is a little bit different from the steepest descent method (gradient method) since step size of the original BP method is always constant η and line search is not carried out. [9] In the original BP method [1], the change of w_{ij} by the pth input pattern is given by:

$$\Delta_p w_{ij} = \eta \, \delta_{ip} \, y_{jp}(w) \tag{5}$$

If the sth layer is an output layer, δ_{ip} of the above equation can be represented as:

$$\delta_{ip} = - f_i'(\bar{z}_{ip}(w))(\bar{x}_{ip} - d_{ip}) \tag{6}$$

If the sth layer is a hidden layer,

$$\delta_{ip} = f_i'(z_{ip}(w)) \sum_k \tilde{\delta}_{kp} \tilde{w}_{ki} \tag{7}$$

where $\tilde{\delta}_{kp}$ and $\tilde{w}_{ki}$ represent the values of δ_{kp} and w_{ki} corresponding to the (s+1)th layer, respectively.

Each weight w_{ij} is updated by the well known back-propagation using the equations (5) to (7). Although the BP method has been considered as one of the most smart algorithms for finding appropriate weights of multi-layered network, it has also several problems. One of the most important problems is the potential for falling into a local minimum of the total error function $E(w)$.* In this paper, we propose a hybrid algorithm which combines the modified back-propagation method (which will be proposed in the next section) with the random optimization method of Solis & Wets [12] in order to find the global minimum of the total error function $E(w)$ in a small number of steps.

III. HYBRID ALGORITHM FOR FINDING THE GLOBAL MINIMUM OF THE TOTAL ERROR FUNCTION OF NEURAL NETWORK

In this section, we propose a new combined algorithm of the modified BP method and the random optimization method [12] in order to find the global minimum of the total error function $E(w)$ in a small number of steps. First, let us briefly refer to the modified BP method and the random optimization method of Solis & Wets [12]

---

* Strictly speaking, the BP method does not ensure monotonous decreasing of $E(w)$. Therefore, even local minimum of $E(w)$ sometimes cannot be found by this method.

3-1. Modified Back-Propagation Method

Recently, Rumelhart and McClelland [13] have proposed the BP method for finding appropriate weights of multi-layered network. This method has attracted great numbers of researchers working in the field of computer, artificial intelligence, and neuro-science. It has been applied to various interesting actual problems.

The basic idea of the original BP method is based upon the steepest descent method which is one of the most simplest optimization algorithms in the field of nonlinear programming. Furthermore, the original BP method uses fixed step size rule which does not perform line search.

As is well known, the conjugate gradient method / the quasi-Newton method [9] exhibit better convergence properties such as n-step quadratic convergence / super-linear convergence than the steepest descent method. Further, the steepest descent method without any line search algorithm cannot ensure convergence even to a local minimum of the objective function.

In the following, we introduce a new modified BP method which exploits the idea of the conjugate gradient method and performs a line search using quadratic polynomial approximation of the total error function in the search direction.

Step 1. Set an initial weight vector $w^{(0)}$ and the parameter values of $\bar{\eta}$, θ, and ϵ. ($\bar{\eta} > 1, 0 < \theta < 1, \epsilon > 0$) Let n be the dimensional number of weight vector w. Let $w = w^{(0)}$ and $k = 0$.

Step 2. Calculate the gradient vector at $w^{(k)}$ of the total error function E(w). Let $g(w^{(k)})$ denote the calculated gradient vector at $w^{(k)}$. This calculation can be done as follows:

$$g(w^{(k)})_{ij} = \frac{\partial E(w)}{\partial w_{ij}}\bigg|_{w=w^{(k)}} = \sum_p \frac{\partial E_p(w)}{\partial w_{ij}}\bigg|_{w=w^{(k)}} = -\sum_p \delta_{ip}\, y_{jp} \qquad (8)$$

, where δ_{ip} is defined by (6) or (7) depending upon whether the sth layer is an output layer or hidden layer.

If $\|g(w^{(k)})\| < \epsilon$, then stop the total calculation. ($\|g(w^{(k)})\|$ denotes the ℓ_2-norm of the vector $g(w^{(k)})$.) Otherwise, let $d_k = -g(w^{(k)})$ (9) and go to Step 3.

Step 3. If $k = 0,\dots,n-1$, then go to Step 4. Otherwise, let $w^{(k-n)} = w^{(k)}$ and $k = k-n$. Go to Step 2.

Step 4. Let $w^{(k+1)} = w^{(k)} + \alpha^* d_k$ (10) , where α^* is determined by the simple line search procedure which will be explained at the Remark 3.R.1.

Further, let $d_{k+1} = - g(w^{(k+1)}) + \beta_k d_k$ (11)

, where $\beta_k = g(w^{(k+1)})^T g(w^{(k+1)}) / g(w^{(k)})^T g(w^{(k)})$ (12)

Let $k = k+1$ and go to Step 3.

<u>Remark 3.R.1:</u>   <u>Line Search in Step 4</u>

The line search procedure utilized in Step 4 consists of a comparatively simple algorithm which determines a minimum of the quadratic polynomial of α which approximates the total error function along the search direction d_k. That is to say, $E(w^{(k)} + \alpha d_k)$ is approximated by some quadratic polynomial function of α. In the following, let us briefly explain this line search algorithm.

Let $^1w^{(k)} = w^{(k)} + d_k$

(i) If $E(^1w^{(k)}) < E(w^{(k)})$, then let $^2w^{(k)} = w^{(k)} + \bar{\eta} d_k$.

    (a) If $E(^2w^{(k)}) \geq E(^1w^{(k)})$, then consider the quadratic polynomial function of $\bar{\eta}$ passing through the three points $(\bar{\eta}=0, E(w^{(k)}))$, $(\bar{\eta}=1, E(^1w^{(k)}))$, $(\bar{\eta}=\bar{\eta}, E(^2w^{(k)}))$. Find $\bar{\eta}^*$ which gives the minimum of the above polynomial function of $\bar{\eta}$. Let $\alpha^* = \bar{\eta}^*$.

    (b) Otherwise, continue the processes $^\ell w^{(k)} = w^{(k)} + \bar{\eta}^{\ell-1} d_k$ ($\ell = 3, 4,$ ...) untill the inequality $E(^\ell w^{(k)}) \cdot \geq E(^{\ell-1} w^{(k)})$ holds. Consider the quadratic polynomial function of $\bar{\eta}$ passing through the last three points $(\bar{\eta}=\bar{\eta}^{\ell-1}, E(^\ell w^{(k)}))$, $(\bar{\eta}=\bar{\eta}^{\ell-2}, E(^{\ell-1}w^{(k)}))$, and $(\bar{\eta}=\bar{\eta}^{\ell-3}, E(^{\ell-2}w^{(k)}))$. Find the parameter $\bar{\eta}^*$ which minimizes this polynomial. Let $\alpha^* = \bar{\eta}^*$.

(ii) If $E(^1w^{(k)}) \geq E(w^{(k)})$, then let $^2w^{(k)} = w^{(k)} + \theta d_k$.

    (a) If $E(^2w^{(k)}) < E(w^{(k)})$, then consider the quadratic polynomial function of θ passing through the three points $(\theta=0, E(w^{(k)}))$, $(\theta=\theta, E(^2w^{(k)}))$, and $(\theta=1, E(^1w^{(k)}))$. Find the parameter θ^* which minimizes this polynomial function of θ. Let $\alpha^* = \theta^*$.

    (b) Otherwise, continue the processes $^\ell w^{(k)} = w^{(k)} + \theta^{\ell-1} d_k$ ($\ell = 3, 4,$ ...) untill the inequality $E(^\ell w^{(k)}) < E(w^{(k)})$ holds. Consider the quadratic polynomial function of θ passing through the last two points and $(\theta=0, E(w^{(k)}))$. Find the parameter θ^* which minimizes this polynomial function. Let $\alpha^* = \theta^*$

Remark 3.R.2:   The above line search has been introduced in order to ensure monotonous decreasing property of the total error function $E(w^{(k)})$.   However, it may turn out that $E(w^{(k)} + \theta d_k) \geq E(w^{(k)})$ for $\theta \in [0,1]$ because perfect line search of θ is not assumed.   Therefore, the process (ii)(b) in Remark 3.R.1 may continue infinitely.   In order to avoid such infinite loop, we have to let $w^{(k+1)} = w^{(k)}$ when we cannot find a parameter w such that $E(w) < E(w^{(k)})$ even after a specified trial number and change the overall algorithm to the random optimization method.

Remark 3.R.3:   Conjugate gradient methods were developed by the strong desire to accelerate slow convergence associated with steepest descent method.   The Fletcher-Reeves method is one of the most familiar conjugate gradient methods and its search direction is determined by (11) and (12).

3-2. Random Optimization Method

In order to prevent the algorithm from stopping on a local minimum of the total error function, we utilize in our hybrid algorithm the nice convergence property of the random optimization method which ensures convergence to a global minimum of the objective function in a compact region.

In 1965, Matyas proposed the following random optimization method[10][11]

RANDOM OPTIMIZATION METHOD OF MATYAS

Step 1.   Select an initial point $w^{(0)}$ in the search domain W.
           Let M be the total number of steps.   Let $k = 0$.

Step 2.   Generate Gaussian random vector $\xi^{(k)}$

           If $w^{(k)} + \xi^{(k)} \in W$, go to Step 3.

           If $w^{(k)} + \xi^{(k)} \notin W$, let $w^{(k+1)} = w^{(k)}$ and go to Step 4.

Step 3.   If $E(w^{(k)} + \xi^{(k)}) < E(w^{(k)})$, let $w^{(k+1)} = w^{(k)} + \xi^{(k)}$.

           If $E(w^{(k)} + \xi^{(k)}) \geq E(w^{(k)})$, let $w^{(k+1)} = w^{(k)}$.

Step 4.   If $k = M$, stop the total calculation.   If $k < M$, let $k = k+1$
           and go to Step 2.

It is well known that this algorithm ensures convergence to a global minimum with probability 1 on a compact set.

In 1981, Solis & Wets proposed the modified random optimization method in order to find the global minimum of the objective function in a small number of steps. Since this method differs from the random optimization method of Matyas only in Step 3, we abbreviate another part of steps.

RANDOM OPTIMIZATION METHOD OF SOLIS & WETS   (Step 3′)

(i)   If $E(w^{(k)} + \xi^{(k)}) < E(w^{(k)})$

let $w^{(k+1)} = w^{(k)} + \xi^{(k)}$ and $b^{(k+1)} = 0.4\xi^{(k)} + 0.2b^{(k)}$.

(ii) If $E(w^{(k)} + \xi^{(k)}) \geq E(w^{(k)})$ and $E(w^{(k)} - \xi^{(k)}) < E(w^{(k)})$,

let $w^{(k+1)} = w^{(k)} - \xi^{(k)}$ and $b^{(k+1)} = b^{(k)} - 0.4\xi^{(k)}$

Otherwise, let $w^{(k+1)} = w^{(k)}$ and $b^{(k+1)} = 0.5b^{(k)}$ ($b^{(0)} = 0$)

It has been reported that the above random optimization method of Solis & Wets exhibits faster convergence than the Matyas' method.   Since this method also ensures convergence to the global minimun of the objective function with probability 1, we shall utilize it as one component of our proposed hybrid algorithm.

3-3. Hybrid Algorithm

We shall propose a new hybrid algorithm that makes use of both the merits of the modified back-propagation method and the random optimization method of Solis & Wets.

Suppose that the objective of the designer of neural network is to decrease the value of the total error function below a small number ε.   Then, the outline of the proposed hybrid algorithm can be described as Figure 1.   Let us explain how it works.   First, parameter training is carried out using the modified back-propagation method.   When the decrease of the value of the total error function E(w) becomes smaller than the specified value $\varepsilon 1 \vee \varepsilon^{*}$, we change the overall descent algorithm from the modified back-propagation method to the random optimization method of Solis & Wets in order to prevent it from falling into a local minimum of E(w).   If the decrease of the total error function becomes larger than $E(w^{(k)})G \vee \varepsilon$ (0 < G < 1), we change the overall descent algorithm from the random optimization method of Solis & Wets to the modified BP method. (Here, $w^{(k)}$ denotes the current weight vector.) This change is performed in order to speed up the overall minimization.** The same procedures are repeated several times.   (Figure 2 demonstrates the basic idea of the proposed hybrid algorithm.)   When the total number of steps exceeds a specified number M, the overall calculation is stopped.

The proposed hybrid algorithm has a nice property that it ensures convergence to a global minimum of the total error function of neural network.   In the following, we shall give the convergence theorem of the proposed hybrid algorithm.

---

* a $\vee$ b denotes   max (a, b)

**The fact that decrease of the total error function has become larger than $E(w^{(k)})G \vee \varepsilon$ means that the current weight vector $w^{(k)}$ has jumped from a local minimum point into a new valley of the total error function.

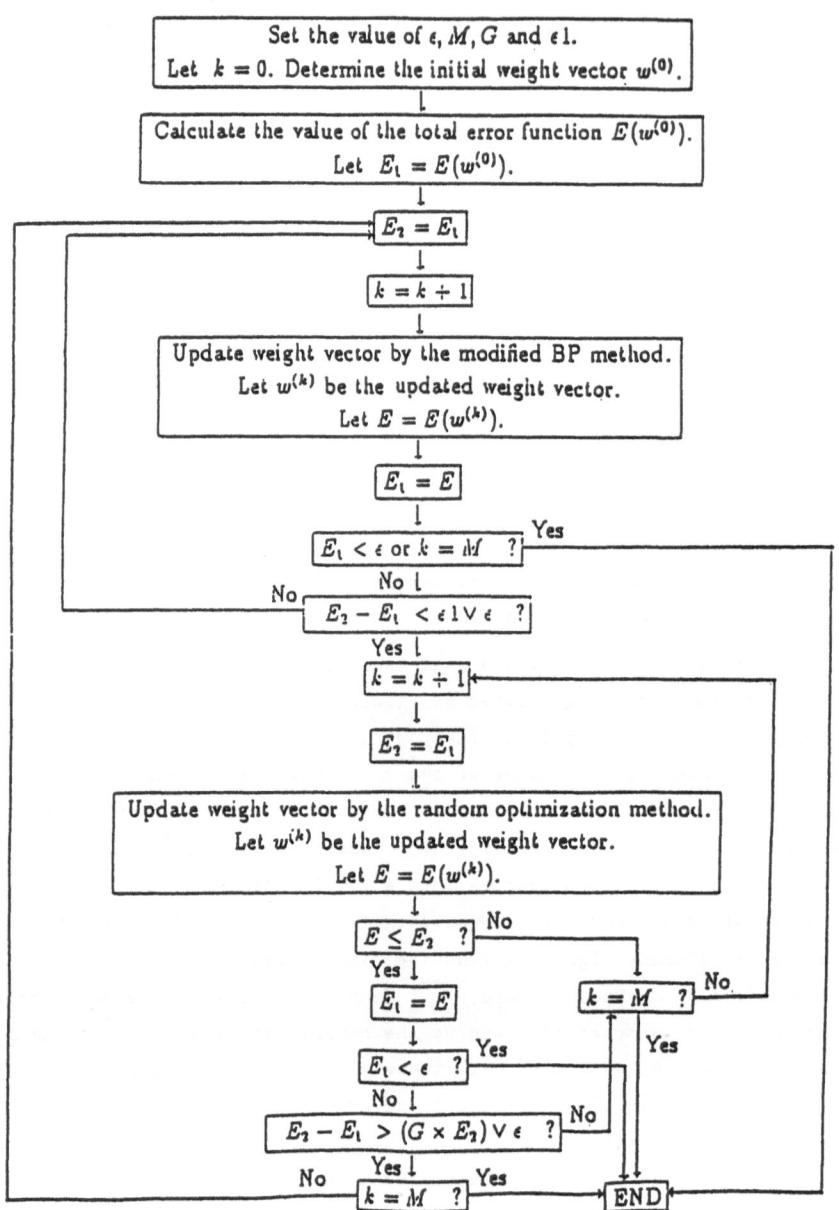

FIGURE 1   Hybrid Algorithm of the Modified Back-propagation Method
and the Random Optimization Method

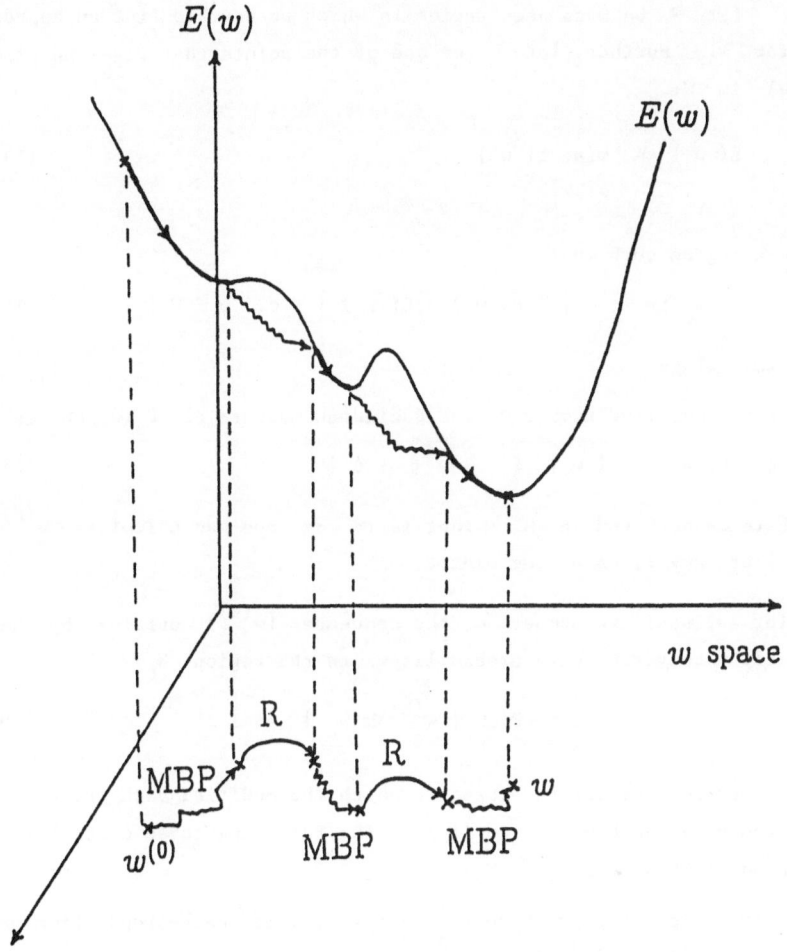

FIGURE 2   Basic idea of the proposed hybrid algorithm   (MBP: Modified back-propagation method   R: Random optimization method of Solis & Wets)

THEOREM: Let W be a compact region in which we have to find an appropriate weight vector w.   Further, let $\hat{w}$ be one of the points that give the global minimum of E(w) in W:

$$E(\hat{w}) \;=\; \min_{w \in W} E(w) \tag{13}$$

Let W_ϵ be a region such that

$$W_\epsilon := \{ w \mid \mid E(w) - E(\hat{w}) \mid < \epsilon, \; w \in W \} \tag{14}$$

Assume the followings:

(1) For any positive number ϵ, the Euclidean measure of $U_\delta(\hat{w}) \cap W$ is positive, where $U_\delta(\hat{w}) = \{ w \mid \parallel w - \hat{w} \parallel < \delta \}$ \hfill (15)

(2) There is no bound on M.   That is to say, one can calculate $w^{(k)}$ (k = 1, ...) as many times as one wants.

Then, for any positive number ϵ, the sequence $\{w^{(k)}\}$ obtained by the proposed hybrid algorithm converges with probability 1 to the region W_ϵ:

$$\lim_{k \to \infty} P\{ \omega \mid w^{(k)} \in W_\epsilon \} = 1 \tag{16}$$

PROOF: Let us consider the cases in which the modified back-propagation method is used in the proposed hybrid algorithm.   The following three cases (1), (2), and (3) correspond to those cases.

(1)   In the first step, we have to use the modified back-propagation method in order to derive $w^{(1)}$ from $w^{(0)}$.

(2)   The preceding calculation of $w^{(1)}$ is done by the modified back-propagation method, and the absolute value of the decrease of the objective function is larger than $\epsilon 1 \vee \epsilon$.

(3)   The preceding calculation of $w^{(1)}$ is done by the random optimization method of Solis & Wets, and the absolute value of the decrease of the objective function is larger than $E(w^{(1)})G \vee \epsilon$.

Let N_1, N_2, and N_3 be the number of times of occurrences of the cases (1), (2), and (3), respectively.   Further let

$$t_1 \epsilon := E(w^{(0)}) - E(\hat{w}) \tag{17}$$

$$h := [t_1] \tag{18}$$

where $[\ t_1\ ]$ means the largest integer which is smaller than t_1.

It is clear that $N_1 = 1$.

From the fact that $E(w^{(k)})$ $(k=0,\dots)$ is a monotonously decreasing function of k, the following inequalities hold:

$$N_i \leq h + 1 \qquad (\ i = 2,3\ )$$

The random optimization method is used whenever the modified BP method is not used. Therefore, the random optimization method is used more than T times to get $w^{(k)}$, where

$$T = k - (\ 1 + 2(h + 1)\ ) = k - (2h + 3)$$

Here, $2h + 3$ is bounded. Therefore, by using the same procedure as used in proving the convergence theorem of the random optimization method [3] one can easily prove the theorem. Q.E.D.

Remark 3.R.4: Since the main objective in the training of the weight vector w in neural network is to decrease the value of the total error function $E(w)$ defined by (4), we have proposed the hybrid algorithm for minimizing this total error function. However, the original BP method consists of the steepest descent method (with fixed step size rule) for minimizing the error function $E_p(w)$ corresponding to each input pattern.

Remark 3.R.5: Since each component of the weight vector w can take arbitrary large values, the assumption that w is included in some compact region seems to be too severe. However, if we confine our calculation to some compact region, for an example, the region in which the absolute values of all the components of the weight vector w are smaller than 10^3, we are surely able to find the global minimum of the total error function $E(w)$ in that compact region by our proposed hybrid algorithm.

Remark 3.R.6: Several parameters such as G, $\varepsilon 1$, and etc. are included in the proposed hybrid algorithm. The selection of the values of these parameters gives significant effect to the efficiency of the overall algorithm.

Remark 3 R.7: In this paper, $\xi^{(k)}$ ($k = 0,1,\ldots$) is assumed to be Gaussian. Therefore, area far from the current point $w^{(k)}$ is chosen only with small probability. On the other hand, area close to the current point $w^{(k)}$ is chosen with high probability. We are now trying to compare the efficiency derived by the uniform distribution of $\xi^{(k)}$ with that derived by the Gaussian distribution of $\xi^{(k)}$. Future research effort should be done in order to investigate what kind of distribution function of $\xi^{(k)}$ is most appropriate to find the global minimum of the total error function $E(w)$ in a comparatively small number of steps.

IV. COMPUTER SIMULATION RESULTS

In this section, computer simulation results of the forecast of air pollution density and the forecast of stock price are presented. The following parameter values of ε, $\varepsilon1$, and M have been chosen: $\varepsilon = 10^{-4}$, $\varepsilon1 = 10^{-3}$, $M = 50000$. As the function $f_i(\cdot)$ in (1), we have used: $f_i(\cdot) = 1/(1 + \exp(-(z-\theta_t)))$

Example 1   Forecast of Air Pollution Density

In this example, we consider prediction of SO_2 density at noon in Tokyo using the informations obtained at 10 a.m. Our objective is to construct a neural network with two hidden layers which emits output "1" (alarm) when the SO_2 density exceeds 5pphm and emits output "0" when the SO_2 density does not exceed 5 pphm. In order to construct such a neural network, we have carried out parameter training using the data obtained over the previous three weeks (October 8, 1970 to October 28, 1970) in order to forecast SO_2 density daily for two weeks in the future. The neural network utilized here has 8 input units which are highly correlated with the SO_2 density at noon in Tokyo (See Figure 3 and Table 1) and one output unit. Table 2 and Table 3 show the learning results of the hybrid algorithm, the BP method, the random optimization method of Solis & Wets, and the modified BP method. In order to demonstrate a specific feature of the learning of each method, Figure 4.1 to Figure 4.4 present the changes of the value of the total error function $E(w)$ when η and variance of ξ assume the typical value 1.

Using the weight vector w having been found by the training procedure, we have obtained forecast of SO_2 density for two weeks. Table 4 shows the success rate of forecasting by each method.

Remark 4.R.1: Table 3 gives an example that the BP method falls into a local minimum of the total error function $E(w)$. On the other hand, the hybrid algorithm and other two methods have successfully found global minimum of $E(w)$. However, Table 4 gives an interesting example in which the success rates of forecasting are nearly the same for each method. In this case, the reduction of the total error function during the learning phase does not result in an improved forecasting.

TABLE 1

Inputs into the Neural Network

x_1: SO_2 density at 10 a.m.

x_2:   (SO_2density at 10 a.m.) −
           (SO_2 density at 7 a.m.)

x_3:   Wind velocity at 10 a.m.

x_4:   Wind velocity at 8 a.m.

x_5: SO_2 density at 9 a.m.

x_6:   (SO_2 density at 9 a.m.) −
           (SO_2 density at 8 a.m.)

x_7: SO_2 density at noon last week

x_8:   (Average SO_2 density between
           8 a.m. and 10 a.m.) − (SO_2
           density at 10 a.m.)

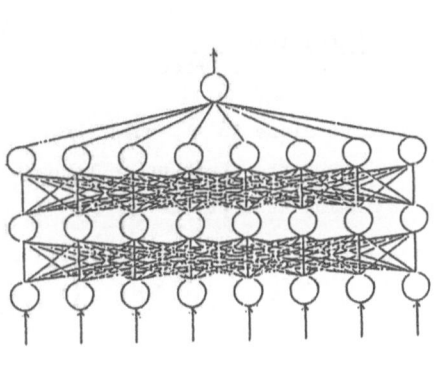

FIGURE 3   Neural Network Used in

Example 1

EXAMPLE 2:   FORECAST OF STOCK PRICE

In this example, we try to give a forecast of stock price by using neural network. The neural network utilized for prediction has two 2 hidden layers, 15 input units, and one output unit.

As inputs into neural network, we have chosen the following 15 variables:

x_1:   Final Stock Price x_2:   Changes of the Stock Price (Today)

x_3:   Turnover (Today) / Averages of the Turnover in the Last Week

x_4:   Turnover (Today) / Averages of the Turnover in the Last Month

x_5:   Changes of the Turnover (Today)

x_6:   The Highest Price in this Year – Final Price (Today)

x_7:   Changes of the Capital of Stockholders x_8:   50 – PER

x_9:   Changes of the Profit of Unit Stock

x_{10}:   Anticipation of the Capital Increase (One Month in the Future)

x_{11}:   Changes of Dow-Jones Averages / Dow-Jones Averages

x_{12}:   Changes of the Currency Rate x_{13}:   Bank Rate

x_{14}:   Pattern of the Changes of the Stock Price

x_{15}: $\dfrac{\text{Final Price – The Least Price in the Last Three Years}}{\text{The Highest Price in the Last Three Years – The Least Price}}$

The objective of the neural network is to emit a value close to "1" when the stock price becomes high and emit a value close to "0" when the stock price becomes low.   Parameter training has been done under the following assumptions:

1)   Stock price of each company is strongly influenced by the current trend.

2)   There are two kinds of trends: a) Increasing Trend (Trend I)   b) Decreasing Trend (Trend D)

We have carried out parameter training of two kinds of neural networks.   As the teacher signal $\bar{d}_i$, we have adopted:

I) Trend I

Let A = $\dfrac{\text{The Highest Stock Price till the End of Next Month – Current Price}}{\text{Current Stock Price}}$

If A > 0.2, $\bar{d}_i$ = 1.   If 0.15 < A $\leq$ 0.2, $\bar{d}_i$ = 0.9.   If 0.10 < A $\leq$ 0.15, $\bar{d}_i$ = 0.8.   If 0.08 < A $\leq$ 0.10, $\bar{d}_i$ = 0.7.   If 0.05 < A $\leq$ 0.08, $\bar{d}_i$ = 0.6. If A $\leq$ 0.05, $\bar{d}_i$ = 0.5.

II) Trend D

Let $B = \dfrac{\text{The Lowest Stock Price till the End of Next Month} \quad - \quad \text{Current Price}}{\text{Current Stock Price}}$

If $B < -0.2$, $\bar{d}_i = 0$. If $-0.2 < B \leq -0.15$, $\bar{d}_i = 0.1$.

If $-0.15 < B \leq -0.10$, $\bar{d}_i = 0.2$.   If $-0.10 < B \leq -0.08$, $\bar{d}_i = 0.3$.

If $-0.08 < B \leq -0.05$, $\bar{d}_i = 0.4$.   If $B > -0.05$, $\bar{d}_i = 0.5$.

Table 5 shows the learning results of the hybrid algorithm when the state of stock price of the mining company N is in the trend I.   Using this neural network, forecasting of the stock price has been carried out.   Table 6 shows this result. All of the outputs from the trained neural network are close to 1 and actual increasing rate of the stock price of this company has become over 20 % in next 2 monthes. This means that the trained neural network makes a good forecast concerning the stock price of the mining company N when the trend is increasing.

   However, several problems emerge:

i) Assume that the current trend is D. Can we still make a good forecast of stock
   price using the neural network having been trained under the assumption that trend
   is also decreasing ?

ii) Assume that the current trend is decreasing (increasing) and we utilize a
   neural network having been trained under the different assumption, that is, the
   current trend is increasing (decreasing).   Can we still make a reasonable forecast of stock price under such an assumption ?

All of our simulation results having been obtained so far give the affirmative answer to the above two problems.   As the examples, we give the following Table 7 and Table 8.   Table 7 shows forecasts of the stock price of the trade company M when the trend is decreasing. (The neural network has been trained under the same assumption.) The very interesting point in this table is that initially outputs are almost zero and increase suddenly at the middle of March, 1990. In fact, the stock price of this company monotonously decreased untill the beginning of April, 1990 and became rather high at the end of April (over 1100 yen).   Therefore, we could consider that this sudden change of the output from neural network suggest the future recovery of the stock price of this company.   Table 8 presents forecasts of the stock price of the mining company N when the trend is increasing.   However, the neural network used for prediction is the one which had been trained under the assumption the trend is decreasing.   In this case, outputs from neural network are not so characteristic. Almost all of them are in the middle part of the interval $(0,1)$.

   From the above two simulations, we observe:

1) The trained neural network emits outputs quite close to the ideal signal when the current trend is the same as that under which neural network has been trained.

2) The trained neural network emits fuzzy outputs when the current trend is different from that under which it had been trained.

However, this observation gives a hint to utilize neural network in dealing stocks:

Let us explain our idea concerning how to use the trained neural networks.

1. Use two kinds of neural networks (having been trained under the two assumptions) in parallel.

2. Input the current data.

3. Watch the outputs from both of the neural networks.

4. Decide Buy, Sell, Stay, and so on.

This idea might be quite helpful. However, in order to verify that, we have to test it for many cases.

EXAMPLE 3: IDENTIFICATION OF CHEMICAL FLOWS IN THE PIPELINE

Our proposed hybrid algorithm for training parameters in a neural network is applied to predict the flow pattern in a pipeline. Reported data on two-phase flow patterns during co-current air-water flow in a horizontal line is used as a sample data. Air and water flow rates in the pipeline are inputs for the network, and the flow pattern name is the output. After the learning process by neural network with some data, flow pattern names for the other data are predicted and compared with the visual observations by human. Although there are some misjudgements, especially in the transitional states, most of the prediction results are identical with human observations. Since we have not enough space, we omit the details.

V. CONCLUSION

A new hybrid algorithm for finding appropriate weights and parameters in the neural network has been proposed. The random optimization method of Solis & Wets has been combined with the modified BP method which exploits the idea of the conjugate gradient method with a simple line search procedure in order to find the global minimum of the total error function in a small number of steps. It has been shown by several examples that the proposed hybrid algorithm is quite useful for finding the global minimum of the total error function of neural network.

However, this fact has been confirmed only for simple example. Therefore, in order to derive general conclusions, future research effort should be directed to investigate the learning property of the proposed hybrid algorithm by applying it to various real problems.

As the algorithms of global minimization, the optimization method using diffusions [7] and the simulated annealing methods [5],[8] are also known. Therefore, in order to find the best tool for training the weights and parameters in the neural network, our effort should also be done to compare carefully the proposed hybrid algorithm with other training methods having been proposed.

ACKNOWLEDGEMENTS

The author would like to thank Mr. T. Totori, Mr. Y. Yoshida, Mr. T. Motoki, Mr. M. Kozaki, and Mr. Y. Mogami for their kind assistance in preparing the manuscript.

REFERENCES

[1] N. Baba, T. Shoman, and Y. Sawaragi, "A Modified Convergence Theorem for a Random Optimization Method", Information Sciences, Vol. 13, pp.159 - pp.166, 1977.

[2] N. Baba, "Convergence of a Random Optimization Method for Constrained Optimization Problems", JOTA, Vol. 33, pp.451 - pp.461, 1981.

[3] N. Baba, "A Hybrid Algorithm for Finding a Global Minimum", Int. J. Control, Vol. 37-5, pp.929 - pp.942, 1983.

[4] N. Baba, "A New Approach for Finding the Global Minimum of Error Function of Neural Networks", Neural Networks, Vol. 2, pp.367 - pp.373, 1989.

[5] V. Cerny, "Thermodynamical Approach to the Travelling Salesman Problems and Efficient Simulation Algorithm", JOTA, Vol. 45, pp.41 - pp.51, 1985.

[6] R. Fletcher and C.M. Reeves, "Function Minimization by Conjugate Gradients", Computer Journal, Vol. 7, pp.149 - pp.154, 1964.

[7] S. Geman & C. Hwang, "Diffusions for Global Optimization", SIAM J. Control and Optimization, Vol. 24, pp.1031 - pp.1043, 1983.

[8] S. Kirkpatrick, C.D. Gelatt, and M.P. Vecchi, "Optimization by Simulated Annealing", IBM Thomas J. Watson Research Center Report, 1982.

[9] D.G. Luenberger, Introduction to Linear & Nonlinear Programming, Addison-Wesley, 1973.

[10] J. Matyas, "Random Optimization", Automation & Remote Control, Vol. 26, pp.246 - pp.253, 1965.

[11] J. Matyas, "Das Zufallig Optimierungs Verfahren und Seine Knnvergenz", Proceedings of the 5th Analogue Computation Meeting, pp.540 - pp.544, 1968.

[12] F.J. Solis & J.B. Wets, "Minimization by Random Search Techniques", Mathematics of Operations Research, Vol. 6, pp.19 - pp.30, 1981.

[13] D.E. Rumelhart and J.L. McClelland, Editors, Parallel Distributed Processing, MIT Press, 1986.

[14] Chua Publishing Company, Various Data on Stock Price in Japan, 1989 and 1990.

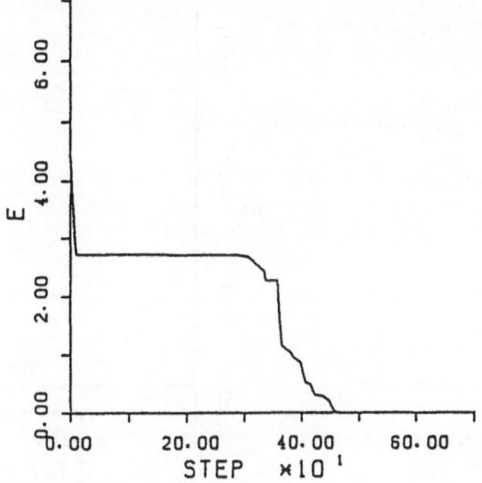

FIGURE 4.1   Changes of the value of the total error function   E(w)
by the hybrid algorithm

($\eta = 1$;   variance of $\xi^{(k)}$ (k=0,1,...) = 1)

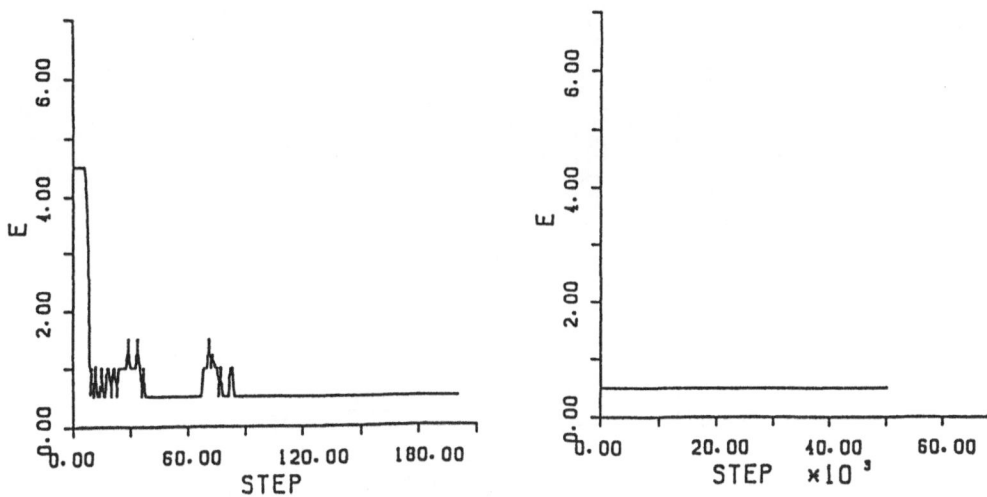

FIGURE 4.2   Changes of the value of the total error function   E(w) ·
by the back-propagation method

($\eta = 1$;   variance of $\xi^{(k)}$ (k=0,1,...) = 1)

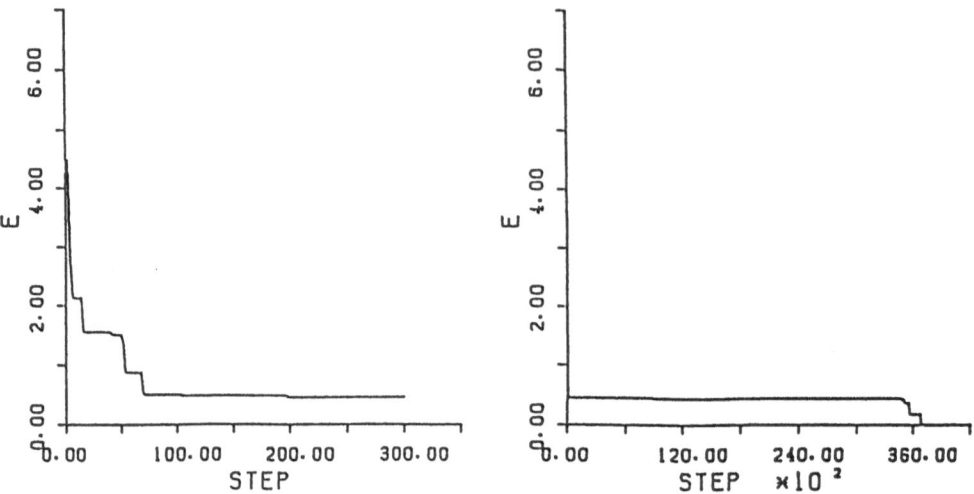

FIGURE 4.3   Changes of the Value of the Total Error Function E(w)

by the Modified Random Optimization Method of Solis & Wets

(η = 1;   variance of $\xi^{(k)}$ (k=0,1,...))

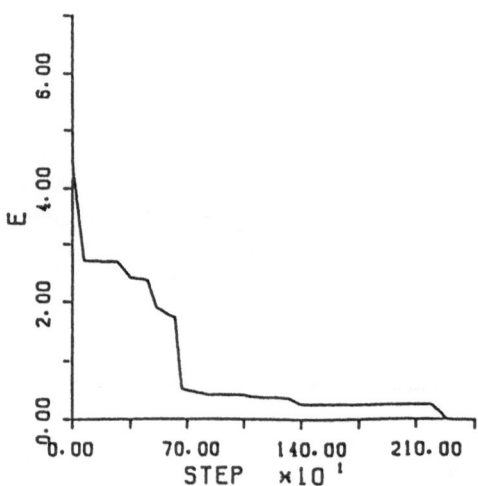

FIGURE 4.4   Changes of the Value of the Total Error Function E(w)

by the Modified Back-propagation Method

(η = 1;   variance of $\xi^{(k)}$ (k=0,1,...) = 1)

TABLE 2

Learning Results by the Hybrid Algorithm (The data from
October 8, 1970 to October 28, 1970 were used for training
the weight vector w.)

$G=0.2$, $\eta=2$, $\theta=0.5$

Hybrid Algorithm			
Variance of ξ	Number of steps	c.p.u.time (seconds)	Final value of the total error function E
0.001	533	1.9800	E < 10^{-4}
0.005	797	2.8965	E < 10^{-4}
0.01	1578	5.6725	E < 10^{-4}
0.05	3886	13.8947	E < 10^{-4}
0.1	5133	18.3223	E < 10^{-4}
1.0	487	1.7655	E < 10^{-4}

$G=0.15$, $\eta=2$, $\theta=0.5$

Hybrid Algorithm			
Variance of ξ	Number of steps	c.p.u.time (seconds)	Final value of the total error function E
0.001	871	3.1548	E < 10^{-4}
0.005	776	2.7973	E < 10^{-4}
0.01	901	3.2589	E < 10^{-4}
0.05	607	2.2120	E < 10^{-4}
0.1	1769	6.3876	E < 10^{-4}
1.0	487	1.7683	E < 10^{-4}

$G=0.1$, $\eta=2$, $\theta=0.5$

Hybrid Algorithm			
Variance of ξ	Number of steps	c.p.u.time (seconds)	Final value of the total error function E
0.001	726	2.6544	E < 10^{-4}
0.005	476	1.7414	E < 10^{-4}
0.01	5864	20.9962	E < 10^{-4}
0.05	534	1.9555	E < 10^{-4}
0.1	451	1.6315	E < 10^{-4}
1.0	487	1.7730	E < 10^{-4}

TABLE 3

Learning Results by the Three Methods (The data from
October 8, 1970 to October 28, 1970 were used for training
the weight vector w.)

Back-propagation Method			
Value of η	Number of steps	c.p.u:time (seconds)	Final value of the total error function E
0.01	50000	305.8416	E=0.49999994
0.1	50000	303.3325	E=0.49999994
0.25	50000	302.8455	E=0.49999994
0.5	50000	302.7549	E=0.49999994
1.0	50000	302.6421	E=0.49999994

Modified Random Optimization Method of Solis & Wets			
Variance of ξ	Number of steps	c.p.u.time (seconds)	Final value of the total error function E
0.001	2048	11.3163	E < 10^{-4}
0.005	4078	24.1137	E < 10^{-4}
0.01	714	4.0434	E < 10^{-4}
0.05	4732	28.3971	E < 10^{-4}
0.1	10468	63.2360	E < 10^{-4}
1.0	38018	230.4277	E < 10^{-4}

Modified Back-propagation Method				
Value of η	value of θ	Number of steps	c.p.u.time (seconds)	Final value of the total error function E
2.0	0.5	2289	8.1441	E < 10^{-4}

TABLE 4

Success Rate of Forecasting by Each Method

A) Hybrid Algorithm

$G = 0.2, \quad \bar{\eta} = 2, \quad \theta = 0.5$

Variance of ξ	0.001	0.005	0.01	0.05	0.1	1.0
Success Rate of Forecasting	13/14	13/14	12/14	13/14	12/14	12/14

$G = 0.15, \quad \bar{\eta} = 2, \quad \theta = 0.5$

Variance of ξ	0.001	0.005	0.01	0.05	0.1	1.0
Success Rate of Forecasting	13/14	14/14	12/14	13/14	12/14	12/14

$G = 0.1, \quad \bar{\eta} = 2, \quad \theta = 0.5$

Variance of ξ	0.001	0.005	0.01	0.05	0.1	1.0
Success Rate of Forecasting	14/14	13/14	11/14	12/14	13/14	12/14

B) Back-propagation Method

Value of η	0.01	0.1	0.25	0.5	1.0
Success Rate of Forecasting	13/14	12/14	12/14	12/14	12/14

C) Random Optimization Method of Solis & Wets

Variance of ξ	0.001	0.005	0.01	0.05	0.1	1.0
Success Rate of Forecasting	13/14	12/14	13/14	10/14	14/14	10/14

D) Modified Back-propagation Method ($\bar{\eta} = 2, \quad \theta = 0.5$)

success rate of forecasting 12/14

TABLE 5

Learning Results by the Hybrid Algorithm

$$(G = 0.1, \ \varepsilon_1 = 10^{-3}, \ \varepsilon = 10^{-4})$$

Variance of	Number of Steps	c.p.u. time (seconds)	Final Value of the Total Error Function
0.001	540	2.2934	0.96914 E-04
0.005	718	3.0729	0.94221 E-04
0.01	152	0.6458	0.94535 E-04
0.05	419	1.7975	0.55700 E-04
0.1	610	2.5627	0.97566 E-04
1.0	1954	8.0545	0.90504 E-04

TABLE 6

Forecast by the Trained Neural Network   (Trend: Increasing)

Date	Final Price	Output	Highest Price After 1 month	Highest Price After 2 Monthes	Rate of Increase	
Nov.1	960	0.99915	1000	1220	4	27
2	953	0.99915	1000	1220	5	28
6	965	0.99915	1000	1220	4	26
7	978	0.99915	1000	1220	2	25
8	983	0.99876	1000	1220	2	24
9	975	0.99894	1000	1220	3	25
10	971	0.99915	1020	1220	5	26
13	964	0.99915	1050	1220	9	27
14	998	0.99915	1050	1220	5	22
15	966	0.90336	1050	1220	9	26
16	970	0.99915	1050	1220	8	26
17	955	0.99915	1070	1220	12	28
20	970	0.99915	1200	1220	24	26
21	970	0.99915	1220	1220	26	26
22	960	0.99915	1220	1220	27	27
24	969	0.99915	1220	1220	26	26
27	1000	0.99915	1220	1220	22	22
28	994	0.90338	1220	1220	23	23
29	973	0.99800	1220	1220	25	25
30	959	0.99915	1220	1220	27	27

TABLE 7

Forecast by the Trained Neural Network (Trend: Decreasing)

Date	Final Price	Output	Lowest Price (After 1 Month)	Rate of Decrease
Mar.1	1070	0.0002	820	23
2	1020	0.0002	820	20
5	1010	0.0004	815	19
6	1020	0.0004	815	20
7	1010	0.0004	815	19
8	1010	0.0004	815	19
9	1010	0.0004	815	19
12	1010	0.0004	815	19
13	995	0.0004	815	18
14	955	0.2489	815	15
15	948	0.2489	815	14
16	940	0.2489	815	13
19	895	0.2489	815	9
20	870	0.2489	815	6
22	845	0.2489	815	4
23	843	0.2489	815	3
26	899	0.2489	815	9
27	910	0.2489	815	10
28	900	0.2489	815	'9
29	890	0.2489	815	8
30	850	0.2489	815	4

TABLE 8

Forecast by Using the Neural Network Which Was Trained at the Period
of Decreasing Trend   (Trend: Increasing)

Date	Final Price	Output	Lowest Price After 1 Month	After 2 Month	Rate of Decrease	
Nov.1	960	0.4039	953	953	1	1
2	953	0.3812	955	955	0	0
6	965	0.3812	955	955	1	1
7	978	0.3812	955	955	2	2
8	983	0.3812	955	955	3	3
9	975	0.3812	955	955	2	2
10	971	0.3812	955	955	2	2
13	964	0.3812	955	955	1	1
14	998	0.3812	955	955	5	5
15	966	0.3812	955	955	1	1
16	970	0.3812	955	955	2	2
17	955	0.3812	959	959	0	0
20	970	0.3812	959	959	1	1
21	970	0.3812	959	959	1	1
22	960	0.3812	959	959	0	0
24	969	0.3812	959	959	1	1
27	1000	0.3812	959	959	4	4
28	994	0.3812	959	959	4	4
29	973	0.3812	959	959	1	1
30	959	0.3812	961	961	0	0

Integrated stochastic approximation program system

Pavel Charamza
Dept. of Statistics, Charles University
Sokolovská 83, 186 00 Praha 8

SUMMARY

The brief description and the full user manual of the Stochastic Approximation program is given. This program contains a large variety of the one dimensional stochastic approximation methods both for the root and the extreme estimates. It can be used also for the root and the extreme estimates of the function of location parameters or the regression quantiles function, in particular. The estimation of parameters of the unknown distribution together with the solution of the LD-50 problem is included as the special part of the program. The program offers the basic statistics and information for comparison of different methods chosen by the user. It was developed for computers compatible with IBM PC and the user needs Turbo Pascal v. 4.0 for its performance.

\- \- \-

1. The list of the methods with the brief description.

The problems solvable by the program are going to be introduced here. They can be divided into several groups. The list of methods that can be used for solving the specified problem are joined after every problem definition. After every method description the list of the parameters that can be defined by the user during the program performance is enclosed. The way how to define these parameters will be explained in chapter 2.

In the mathematical formulation of the problems we shall use the following notation:

(R, B) ... one-dimensional Euclidean measurable space with Borel σ-algebra,

E ... expectation functional on some probability space (Ω, A, P),

$X(t)$ ... the value of the estimator at time t,

$Y(x)$ ... the value of the measurement at the point x (this value can be generated by the random numbers generator).

A/Problem of finding the root of the regression function

The problem can be formulated as solving the equation

$$Eg(x,.)=\alpha, \qquad\qquad /1/$$

where $g:(Rx\Omega,B\otimes A)\longrightarrow(R,B)$ and α is some given real number which can be chosen arbitrarily in the program. The function g can also be defined by user as will be shown in more details in chapter 2.

Methods available for solving of problem /1/:

A-i/ Robbins-Monro procedure

The standard scheme

$$X(t+1)=X(t)+\frac{a}{t}(Y(X(t))-\alpha)$$

is used for solving /1/. See e.g. Nevel'son,Khasminskij [1972] for the basic asymptotic properties of the sequence $X(t)$.

Parameters: a,α.

A-ii/ Adaptive Robbins-Monro procedure

For finding the solution of /1/ with $\alpha=0$ the adaptive scheme

$$X(t+1)=X(t)- \quad (Y(X(t)-\frac{c}{t^\gamma}) + Y(X(t)+\frac{c}{t^\gamma}))/(2*t*W(t)),$$

is used, where

$$W(0)=0,$$
$$W(t+1)=max[max(r1,min(r2,W'(t+1)))],$$
$$W'(t+1)=W(t)+[(Y(X(t)+\frac{c}{t^\gamma}) - Y(X(t)-\frac{c}{t^\gamma}))/(2*c*t^\gamma)-W'(t)]/t.$$

The scheme was taken from Dupac[1981].

Parameters: c,r1,r2,γ.

A-iii/ Procedure based on isotonic regression

The Mukerjee's algorithm (Mukerjee[1981]) is used for solving /1/. It is characterized by the fact that our measurements can be taken only at the point of the lattice L which is defined as the $L\overset{def}{=}\{a+ld;l$ integer$\}$, where a,d are given real numbers. The idea of the algorithm is to find the root of the isotonic function fitted (in the least squares sense) by the data gained up to time t. By isotonic fitting we mean the nondecreasing function f which minimizes

$$\sum_{t=1}^{n} (Y(X(t)) - f(X(t)))^2. \qquad\qquad /2/$$

The measurements (or simulations) at time t+1 are provided at the points of the lattice L that are the nearest to the estimated root.

See Mukerjee [1981] for more details and convergence theorem. The theory about the least square isotonic regression including the effective algorithm can be found in the book Barlow, Bartholomew, et. al. [1972].

Parameters: d (the step of the lattice L), α.

A-iv/_Procedure_based_on_quasiisotonic_regression

The modification of the Mukerjee's algorithm was suggested by Dupac [1987]. He considered the same algorithm where he used the root of the quasiisotonic fit instead of the isotonic one. The quasiisotonic function is defined in his paper as the function f for which there exists θ such that $f(x) \leq α$ if $x \leq θ$ and $f(x) \geq α$ if $x \geq θ$. The easy algorithm for finding quasiisotonic fit and the convergence theorem are given in Dupac's paper too. In his algorithm the measurements can be made only at the points of the lattice L as in the Mukerjee's case.

Parameters: d (the step of the lattice L), α.

A-v/_Procedure_based_on_L1-quasiisotonic_regression

The relations between the algorithm of Mukerjee and that one of Dupac lead to the L1-quasiisotonic regression which give almost the same results as the Mukerjee's method but the algorithm for finding the root is that type of Dupac. The theoretical background concerning the relations among these three methods and their modifications can be found in Charamza [1989].

Parameters: d (the step of the lattice L), α.

A-vi/_Pflug_procedure

The procedure is based on the algorithm

$$X(t+1) = X(t) - a[Y(X(t) - α],$$

where a changes adaptively according to the rule of Pflug [1988]. This rule is based on the so called 'oscilation test', i.e. when the sequence of X(t) is stylistically proved to oscillate around some value, the value of a is multiplied by $γ_2$, which should be less than 1. The oscillation test is as follows. Let us denote the time when the a was changed last time by τ and the point which solves /1/ by x^*. Now we find the first T for which

$$\frac{1}{T} \sum_{i=1}^{T} Y(X(τ+i+1)) Y(X(τ+i)) \leq a \, σ^2 \, f' \left\{ \frac{f'}{2} - \frac{a(f')^2}{2} - 1 \right\} + γ_1,$$

where $σ^2$ is the variance of the error at the point x^*, $f' \overset{def}{=} \frac{\partial}{\partial x} Eg(x^*, ω)$

and γ_1 is some constant. At the time T the parameter a can be changed described. In practice the user need not know the exact values σ^2, f' and he must estimate them (see Pflug [1988]). Nevertheless, we use the exact values here that should be given before running the algorithm. *Parameters:* a (starting value), σ^2, f', γ_1, γ_2, α.

B/Problem_of_finding_the_root_of_the_regression_quantile_function

Procedures of this paragraph solves iteratively the equation

$$Q(x, \gamma) = \alpha, \qquad /3/$$

where $Q(x, \gamma)$ is defined as the γ-quantile of the distribution according to which the observations at the point x are obtained.

B-i/ Procedure_based_on_isotonic_quantile_regression

The idea of the Mukerjee's algorithm (see A-iii) can be spread for solving the problem /3/ if we take the least absolute deviations isotonic fit instead of the least squares one, i.e. we solve

$$\sum_{t=1}^{n} |Y(X(t)) - f(X(t))|$$

instead of /2/. The algorithm for finding the least absolute deviations isotonic regression which is analogous to the least squares one was suggested by Menendez, Salvador [1987]. The convergence results concerning the stochastic approximation procedure based on this fit were proved in Charamza [1989].

Parameters: d (the step of the lattice L), α, γ.

Remark 1: The estimates of the γ-quantiles (at each point at which we made observations up to time n) are necessary for obtaining the least absolute deviations isotonic fit. These estimates are computed according to the recursive M-estimate procedure which is as follows:

Recursive M-estimate procedure:

Recursive M-estimate $\tilde{\theta}_{i+1}(\psi)$ based on the observations $X_1, \ldots, X_{i+1}$ is defined as

$$\tilde{\theta}_{i+1}(\psi) = \tilde{\theta}_i(\psi) - \tilde{\gamma}_i(\psi)(i+1)^{-1}\psi(X_{i+1}, \tilde{\theta}_i(\psi)), i \geq 0, \qquad /4/$$

where

$$\tilde{\gamma}_i(\psi) = \tilde{\gamma}_i^*(\psi) \qquad \text{if } i^{-\delta} \leq |\tilde{\gamma}_i^*(\psi)| \leq \ln i$$

$$= i^{-\delta} \qquad \text{if } |\tilde{\gamma}_i^*(\psi)| < i^{-\delta}$$

$$= \ln i \qquad \text{if } |\tilde{\gamma}_i^*(\psi)| > \ln i$$

and

$$\tilde{\gamma}_i^* = (2ti)^{-1} \sum_{j=1}^{i} \left[\psi(X_j, \tilde{\theta}_{j-1}(\psi) + j^{-\beta}) - \psi(X_j, \tilde{\theta}_{j-1}(\psi) - tj^{-\beta}) \right] j^{\beta}.$$

The initial condition is computed as the sample quantile or trimmed mean from s values. The values of s and of the constants necessary for computing the recursive M-estimate can be changed in the program as will be explained in 2.3.4, paragraph *M-estim. parameters*. For convergence results concerning the recursive M-estimates see e.g. Huskova [1988]. For the purpose of estimating quantiles the function ψ is taken as $\psi(x) = \gamma \ I(x \leq 0) + (1-\gamma)I(x \geq 0)$, where I is characteristic function of a set.

B-ii/ Procedure based on quasiisotonic quantile regression

Here we took least absolute deviation quasiisotonic regression instead of the least square one in the Dupac's algorithm (see A-iv). Taking the root of this LAD fit and providing next observations at the points of the lattice L that are the nearest to this root we get the sequence which was proved to converge to the solution of /3/ in Charamza [1989]. The estimates of the quantiles that need to be compute for finding LAD quasiisotonic fit are evaluated according to the recursive M-estimate algorithm from Remark 1, B-i.
Parameters: d (the step of the lattice L), α, γ.

B-iii/ Derman procedure

For finding the solution of /3/ the generalization of Derman procedure (Derman [1957]) is used. This generalization can be found and proved to converge in Charamza [1984] and is as follows.
Algorithm 1: Provide the observation at time t on the level X(t), where

$$X(1) \in L \ \text{arbitrary}$$

$$X(t+1) = X(t) + d \ \text{with probability} \ \frac{1}{2\gamma} \ , \ \text{if} \ Y(X(t)) \geq \alpha$$

$$X(t+1) = X(t) - d \ \text{with probability} \ 1 - \frac{1}{2\gamma} \ , \ \text{if} \ Y(X(t)) \geq \alpha$$

$$X(t+1) = X(t) + d \ \text{with probability} \ 1, \ \text{if} \ Y(X(t)) < \alpha.$$

- - -

We define estimate of X^* at time t as θ_t which is the element of

$$\underset{i \in L}{\operatorname{argmax}} \sum_{t=1}^{\infty} I\{X(t)=i\}.$$

We take θ_t as the arithmetic mean is there exist more solutions. Similarly as in the case of isotonic and quasiisotonic regressions (see A-iii,A-iv,B-i,B-ii) the observations can be provided only at the points of the lattice L.
Parameters: d (the step of the lattice L), α, γ.

B-iv/ Procedure_based_on_isotonic_regression

The problem of solving /3/ can be transformed using the observations Z(X(t)),

$$Z(X(t))\overset{def}{=}I[Y(X(t)\geq\alpha], \qquad\qquad /5/$$

instead of Y(X(t)). After this transformation we look now for the solution of equation

$$E(Z(x)) = 1-\gamma, \qquad\qquad /6/$$

which is of the type of /1/. Therefore the methods from the paragraph A can be used for solving /3/. Here we used the method A-iii.
Parameters: d (the step of the lattice L), α, γ.

B-v/ Blum_procedure

The modification of the Robbins Monro method suggested by Blum [1954] can be used for solving /3/. The transformation /5/ as in B-iv is being evaluated in order to transfer /3/ to /1/. After this transformation the Robbins-Monro procedure (viz A-i) is used for solving /1/.
Parameters: a, α, γ.

B-vi/ Bather_procedure

We assume in this procedure that $\gamma=1/2$ and $\alpha=0$. Using transformation /5/ we have random variables Z(.) instead of Y(.). These variables can achieve only 0 or 1 value. Although the function EZ(x) is not increasing in x the algorithms for finding the solution of so called LD-50 problem can be used for our original one. (See paragraph F how the LD-50 problem is defined.) Here we use the Bather method for solving LD-50 which is of the form

$$X(t+1) = \bar{X}(t) - \frac{a}{t^\gamma} (2*S(t)-t)$$

where $\bar{X}(t)$ is the average of the points at that we made observations up to time t and S(t) is the sum of the values of Z(X(t)) up to time t. See Bather [1989] for more details and discussion over this type of methods.

Parameters: a, γ.

C/Problem of finding the root of the regression location parameter function

Let us denote by F_x the distribution function of the observations provided at the point x. Let $H:(R^2,B^2)\longrightarrow(R,B)$. By the regression location parameter function we shall mean the real measurable function m(x) which is defined as

$$m(x) \stackrel{\mathrm{def}}{=} \underset{\theta}{\mathrm{argmin}} \int H(y,\theta)\ dF_x(y).\qquad\qquad /7/$$

The regression function or the regression quantile function are the special cases of /7/ choosing $H(y,\theta)=(y-\theta)^2$ or $H(y,\theta)=|y-\theta|$. The methods listed below are proved to solve iteratively the equation

$$m(x)=\alpha.\qquad\qquad /8/$$

C-i/ Procedure based on isotonic regression

The solution of /7/ can be obtained using the algorithm of Charamza [1989] which is of the type of Mukerjee. He used for estimating the solution of /8/ the root of the "robust isotonic regression" which is defined as the isotonic function f which minimizes

$$\sum_{t=1}^{n} H(Y(X(t)),f(X(t))).$$

The recursive M-estimator from Remark 1 (see B-i) is used for the location parameters estimates that are necessary for evaluation of this robust regression. The function ψ from this recursive estimator plays the role of the partial derivative of the function H according the variable θ. The functions H and ψ have to be defined by user. The way how to define them is described in the chapter 2, paragraph 2.1.2. *Parameters:* d (the step of the lattice L), α, γ.

C-ii/ Procedure based on quasiisotonic regression

This algorithm was suggested and proved to converge under some circumstances in Charamza [1989]. It is of the form as in C-i, where we take quasiisotonic fitting instead of the isotonic one. In fact, methods from paragraphs C-i, C-ii are generalizations of the Mukerjee's and Dupac's approaches (see A-iii, A-iv). As in the C-i case, the function ψ plays the role of the partial derivative of the function H according to variable θ. These functions should be defined by user before running the program as will be shown in the chapter 2, paragraph

2.1.2.

Parameters: d (the step of the lattice L), α, γ.

D/Estimation of extreme of regression function

The methods from this paragraph are designed for finding the extreme of the function $m(x) \overset{def}{=} Eg(x, .)$, where the function g is the same as in /1/. The problem can be written as

$$m(x) \longrightarrow \min! \tag{/9/}$$

D-i/ Kiefer-Wolfowitz method

The standard basic algorithm (see e.g. Dupac [1984])

$$X(t+1) = X(t) - \frac{a}{t^{5/6}} Z(t),$$

where

$$Z(t) = \left[Y(X(t)+c/\sqrt[6]{t}) - Y(X(t)-c/\sqrt[6]{t}) \right] \frac{1}{2c}.$$

is used for solving /9/.

Parameters: a, c.

D-ii/ Isotonic regression for differences

This procedure solves /9/ on the discrete lattice L (see A-iii). Using the transformation

$$Z(X(t)) = Y(X(t)) - Y(X(t)-d) \tag{/10/}$$

(where d is the step of the lattice L) the problem /9/ can be transferred to /1/ with $\alpha = 0$. For solving /1/ the isotonic regression procedure A-iii is used.

Parameters: d (the step of the lattice L).

D-iii/ Quasiisotonic regression for differences

Using the transformation /10/ the problem /9/ is transformed to /1/ with $\alpha = 0$ as in D-ii. For solving the problem /1/ the procedure A-iv is used.

Parameters: d (the step of the lattice L)..

E/Estimation of quantile

Procedures of this paragraph can be used for finding the γ-quantile of an unknown distribution having the random sample from it. All the methods are recursive and based on the ideas from paragraph B using the following arguments:

We denote the value of the t-th variable from our random sample by

Z_t and by F the distribution of this variable. Let us consider the function $m(x)=-F^{-1}(\gamma)+x$. Its root is clearly equal to $F^{-1}(\gamma)$. Let we further take the value $Y(X(t))\overset{def}{=}X(t)-Z_t$ as our observation at the point $X(t)$. The $(1-\gamma)$-quantile of the distribution of the variable $Y(x)=x-Z_1$ is equal to $m(x)$ (if F is increasing function). This is implied by

$$P[Y(x)\le m(x)]=P[Z_1\ge F^{-1}(\gamma)]=1-\gamma.$$

The problem of finding the γ-quantile of unknown distribution is thus transferred to the problem of finding the root of the regression $(1-\gamma)$-quantile function.

The previous arguments show how to use the methods from paragraph B for estimating unknown γ-quantile. Nevertheless there are some advantages in using the procedures from this paragraph instead of those from B. We can prepare our random sample on the disk file in advance. How to handle the data in this case will be described in the chapter 2, paragraph 2.2.1 and also 2.3.3.

E-i/Tierney method

The idea of the procedure is according to Blum [1954] (see B-v). See also Tierney [1983] for some investigations concerning this method. The recursive formula is taken as

$$X(t+1)=X(t)-a/(t+1)*(I(Y(t+1)\le X(t))-\gamma).$$

Paerameters: a,γ.

E-ii/ Procedure based on quasiisotonic regression

The procedure B-ii is used for the problem /3/.
Parameters: d(the step for the lattice L), γ.

iii/ Procedure based on isotonic regression

The procedure B-i is used for solving /3/ with $\alpha=0$.
Parameters: d (the step for the lattice L), γ.

iv/ Derman procedure

The procedure B-iii is used for solving /3/ with $\alpha=0$.
Parameters: d (the step of the lattice L), γ.

F/LD-50 problem

Let us have the unknown distribution function F. The problem is to find the γ-quantile of this distribution. In comparison with the problem formulated in the previous paragraph we have not the random sample from this distribution to our dispose. The only information that we obtain about this distribution is given by independent random variables Y(X(t)) distributed according the alternative law:

$$Y(X(t))=0 \quad \text{with probability } 1-F(X(t))$$
$$Y(X(t))=1 \quad \text{with probability } F(X(t)).$$

The mean value of the variable $Y(x)$ is equal to $F(x)$. Therefore the methods from paragraph A with $\alpha=\gamma$ could be used for finding the unknown γ-quantile. Nevertheless two methods for LD-50 problem were introduced in the program separately. The first of them is the method A-iii with parameter α predestined as 0.5. The second one is of the different type.

F-i/ Procedure based on the isotonic regression

The isotonic regression method (see A-iii) is used for LD-50 problem.
Parameters: d (the step of the lattice L).

F-ii/ Derman procedure

The procedure solves the general case of finding the γ-quantile having observed the 0-1 variables. It is based on algorithm 1 from B-iii, where we put $\alpha=0.5$. Recall that the observations can be made only at the points of the lattice L in this procedure.
Parameters: d(the step of the lattice L),γ.

- - -

2. Running the program

At first let us give the list of the files stored on the Distribution diskette. We recommend the user to copy them to the special directory on his hard disk. The user will also need to copy GRAPH.TPU unit and *.BGI driver from his Turbo Pascal v.4.0 environment to this directory, too. If he has CGA monitor he should use CGA.BGI, use HERC.BGI for Hercules monitors and EGAVGA.BGI for EGA or VGA monitors.
Files on the diskette: STAPR1.PAS, MAIN.TPU, GLOBAL.TPU, MCRT.TPU, GRAFIKA.TPU, UTILITY.TPU, RANDGEN2.TPU, FUNKCE.TPU, USER.PAS, STAPR1.EXE.

2.1 Before running the program

The program gives the user a possibility to solve the problems from the chapter 1 both for simulated data or data obtained in practice. In the case of the simulation, values of measurements at the point x are evaluated according to the function $g(x,\omega)$, where ω is a random variable with the distribution selected by user (see paragraph 2.4 for

how to choose distribution of ω). Function g can be programmed by user
as the Turbo Pascal v 4.0 function. This function should be linked to
the main program at first. Now we describe the way how to do this.

2.1.1 _Definition_of_the_function_g(x,ω)

Let us suppose that the function g is of the form $g(x,\omega)=x*\omega+\omega+x^3$.
By any editor the user is familiar with he can create the file
USER.PAS as follows:

unit user;
interface
 function g(x,omega:real):real;
implementation
 function g;
 begin
  g:=x*omega+omega+x*x*x;
 end;
end.

The bold-face part is obligatory! The normal-face text is up to the
user choice under the only condition - it must be the correct
expression in the Turbo Pascal version 4.0 language.

- - -

The second thing that should be done before running the program is
to define the functions $H(y,\theta)$ and $\psi(y,\theta)$ that have the meaning
explained briefly in chapter 1, paragraph C-i. These definitions are
necessary only in the case when the user wants to use procedures C-i or
C-ii for solving his problem. In this case the file USER.PAS will take
the following form:

2.1.2 _Definition_of_the_functions_H(y,θ)_and_ψ(y,θ)

unit user;
interface
 function g(x,omega:real):real;
 function H(y,theta:real):real;
 function PSI(y,theta:real):real;
implementation
 function g;
 begin
  g:=x*omega+omega+x*x*x;
 end;

```
function H;
begin
  H:=abs(y-theta);
end;
function PSI;
begin
  if theta - y≥0 then PSI:=1
                 else PSI:=-1;
end;
end.
```

The bold-face text is obligatory as in the previous case. For our example we took absolute value as the function H. Reminding C-i, ψ is the derivative of H according to the second variable. From here the definition of Pascal function PSI followed.

Remark: For the definition of g,H,ψ the following functions can be used together with the standard Turbo Pascal ones:

- min(x,y:real):real;{returns the lower number from x,y}
- max(x,y:real):real;{returns the larger number from x,y}
- sign(x:real):shortint;{function sign(x)}
- Power(x,y:real):real;{returns x^y}

If the user wants to use any of these functions he must change the interface part of the USER.PAS file as follows:

```
unit user;
interface
  uses utility;
  function g(x,omega:real):real;
  function H(y,theta:real):real;
  function PSI(y,theta:real):real;
```

continuing in the same manner as in the previous examples.

2.1.3 Compiling, linking and running program

Now the user has to compile and link USER.PAS file to the main program. He will need the Turbo Pascal version 4.0 to do this and should provide following operations:

Load the USER.PAS file into the Turbo Pascal environment. Compile USER.PAS to disk. {Select the *Disk* directive in Pascal Compile menu and press Ctrl-F9. See Turbo Pascal manual for details}. If succesfull then load STAPR1.PAS into the Turbo Pascal environment and press F9. Then run the program pressing Alt-R.

− − −

Attention 1: In the moment of linking the program (by pressing the F9 key) the GRAPH.TPU file, distributed together with Turbo Pascal version 4.0, must be present in the active directory. When the user runs the program the graphics driver relevant to his graphics card must be in the same directory. For the suitable graphics driver for his computer the user can consult with his Turbo Pascal manual.

− − −

The title of the program with its number of version should appear on the screen now. If not, there is some problem with the graphics driver. When O.K. the user can press any key to activate the main menu with commands *Provide, Input, Generator, Output.* Now we give the description for each of the main menu command and its subcommands. The way how to move through the menu structure is given in paragraph 2.7.

2.2 Provide

Under this command there is the sub menu with three items: *Method, Comparison, Quit.*

2.2.1 Method

Under this command the user can select the desired procedure for solving his problem. The procedures from the chapter 1 can be divided in the three main groups. Methods for solving /1/ (introduced in the paragraphs A,B,C), methods for solving /9/ (see paragraph D) and finally the methods for estimation of quantiles. Having chosen the Method command the user can select one of these three groups from the sub menu with the items *Root estimation, Extreme estimation* and *Parameter estimation.*

Root estimation

Under this command the user determine whether he prefers the method from paragraph A (item *Regression function*), method from paragraph B (item *Quantile regression*) or from paragraph C (item *M-estimate function*).

Extreme estimation

The only item *Regression function* is given under this sub command.

Quantile estimation

The two items menu (*Quantile estimation, LD-50*) gives the possibility to choose between procedures from paragraph E, F respectively. After the selection of the item wanted the full list of the methods from the relevant paragraph is offered. Choosing the desired procedure the windows with predestined values of the parameters appears on the screen. The user can change any value of the introduced parameters using edit instructions (see 2.6).

Brief help about chosen procedure is given at the bottom of the screen. (In the case of procedures from paragraph E there is one special circumstance which doesn't appear for other procedures. In this case the input data doesn't depend on the place where the measurements are provided. The input is a random sample. If the user select *Real data* evaluation (see paragraph 2.3.3) the program gives him the possibility to select whether the data will be taken from the keyboard or from the disk file. Thus if he chooses one of the procedures from paragraph E the sub menu with items *Keyboard, Disk* appears on the screen before the parameters selection.) After selecting parameters by pressing Esc key (or Return in the case of one parameter) the full information about the parameters chosen is given including the information about the distribution of ω. The user can see also the information concerning the stopping time and the initial value for the method. These values can be selected under the *Input* command as will be described in paragraph 2.3. The distribution of ω can be chosen in the *Generator* menu (see 2.4). If the user is agreed with the parameters he can press 'y' and the selected method will be provided (see paragraph 2.5 for the possibilities of how the results will be performed). If he is not agreed he may press any other key to return to the main menu.

The logical structure of the menus under the Method command can be illustrated by the TREE 1.

2.2.2 Comparison

The results of the last five methods that had been computed were stored on diskette (or disk) for future comparisons. If more than five methods had been provided the first four and the last one were preserved. Under *Comparison* command from the *Provide* menu the user can select those methods he wants to compare together. Entering this command the sub menu with the following items appears: aLl, Method A, Method B, ...,eRase. Length of the sub menu depends on the number of methods that have been provided under the Method command. On the right hand sight of the screen the description of parameters of the method prepared for comparisons is given. The user can select the desired methods for comparisons by pressing Enter on the method items chosen or by pressing the letters denoting them. The selected methods are highlighted. The user can select all the methods at once using the *aLl*

Method

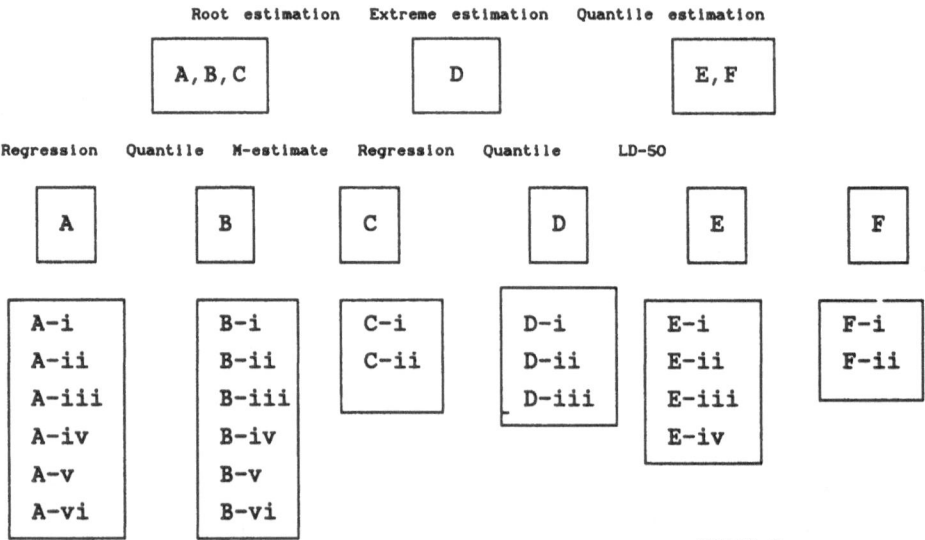

TREE 1

command. The unselecting of the methods is provided in the same manner, i.e. by pressing Enter on the item of the already selected method or by pressing the letter denoting this method. After the selection the user can provide comparisons according to the options established in the *Output* menu (see paragraph 2.5) by pressing Esc-key. On the other hand he can erase the selected methods from the list using the *eRase* command from the menu. This enables him to prepare place in the stack for other methods.

2.2.3 Quit

The user can terminate the program using this option. The question 'O.K. to quit? (Y/N)' appears under this item. By pressing 'y' the user terminates the program. If he presses another key the main menu appears again.

The second command from the main menu is

2.3 Input

Using this command the user can enter the input information for the methods. The following items appears on the sub menu under the *Input* command: *Number of iterations, Stopping time, sTarting value, Input, input Data file, M-estim. parameters.*

2.3.1 Number of iterations and Stopping time

Each method is provided from time 0 to Stoptime which can be entered under the *Stopping time* command using the edit instructions (see 2.6). This whole process can be repeated x times where x is entered under the *Number of iterations* command. This enables to get an information about the sample asymptotic variance of each method. This information is given in the numerical form of results (see paragraph 2.5.2). See paragraph 2.5 for how to choose this type of output.

2.3.2 sTarting value

Under the *sTarting value* command there is the sub menu containing two items: *Deterministic* and *Random*. Choosing the *Deterministic* item the user defines the initial condition X(0) for the selected method as some constant value. He can change the predestined starting value following editing instructions (see 2.6). If he chooses *Random* item the initial value X(0) will be generated as a normally distributed random variable. In this case each of the provided method will begin from the different point. The parameters of the normal distribution can be changed under the *Random* command using edit instructions again.

2.3.3 Input and input Data file

As was mentioned at the beginning of this chapter the program gives the user the possibility to solve the problems from the chapter 1 both for simulated data or data obtained in practice. Under the Input command the user chooses what possibility he prefers for his problem. There are two items under this command: *Simulations* and *Real problem*. Choosing the *Simulations* item the methods will be provided automatically using the definition of the function g and of the distribution ω that is defined under the *Generator* command (see 2.4). In the second case the method will be interrupted at every time moment asking the user for the value of the measurement or it will take data from the data file without interruption. The data can be taken from the data file only in the situation when their values don't depend on the places where the measurements are provided. This situation arises in the quantile estimation based on the random sample that means when the procedures from paragraph E of chapter 1 are considered. When the user choose *Real problem* item then after selecting one of the methods from this paragraph he is asked whether the input will be from *Keyboard* or from *Disk* (see also paragraph 2.2.1). Selecting the *Disk* item the data are read from the file which name should be entered under the *input Data file* command.

2.3.4 M-estim. parameter

There are some methods in this program (see chapter 1, paragraphs B-i, B-ii, C-i, C-ii, E-ii, E-iii) evaluation of that works with recursive M-estimator. The definition of this estimator is given in Remark 1, see B-i, and depends on the parameters $t \neq 0$, $0 < \beta < 1/2$, $0 < \delta < 1/2$, $\tilde{\gamma}_0 \neq 0$, s and $\tilde{\theta}_0(\psi)$. The values of these parameters can be changed using the *M-estim. parameters* command following the edit instructions (see 2.6).

The third command from the main menu is

2.4 Generator

Using this command the user can select the distribution of ω which the evaluation of the function g depends on (see paragraph Before running the program for further details). There are five items under this commands: *Uniform, Normal, Cauchy, Double exp.* and *Seed*. Choosing one of the first four items the user determines the distribution of ω which will be taken into account if the *Simulation* command will be selected under the *Input* item (see 2.3.3). The values of ω are used for evaluation of the function g. If there are more than one ω the user needs for this definition (e.g. $g(x, \omega) = x + x \ast \omega + \omega$) then the independent numbers are generated from the underlying distribution. After choosing the distribution the user can change the parameters of it following edit instructions (see 2.6). The random generator composed from three congruent generators was taken from Wichmann, Hill [1987]. The user can control the seeds of this generator by choosing the *Seed* item from the *Generator* menu. He can also change any of three numbers that define the random generator following editing instructions again. The standard setting (s1=1, s2=10000, s3=3000) is set by the program before starting any of the methods if no change in the seeds has been done.

The last command from the main menu is

2.5 Output

There are six items under this command: *Numerical, Graphical, Disk, First time, Time interval, Output data file*. Using one of the first three items the user can select the type of output he needs. This type of output will be performed either after each method evaluation (see 2.2.1) or if the user compares methods already evaluated (see 2.2.2). There are two options under each of these items: *On, oFf*. If *On* option is specified for some type of output the results of specified method or comparisons will be resumed in this form. Setting all three items to *On* then all the three types of output are provided. The user can also set all these items to *oFf*. This can be useful in the moment when he needs some comparisons only. He can switch off the output when the methods are computed and switch the desired output on when the comparisons will be provided. If the methods with a long path are computed or compared then the user usually needs to study the results only at some time moments. Sometimes he can be interested only in the behavior of the end of the path. In the program he can define the first time moment from which the path is shown together with the time moments at that the values of procedure(s) are given. This all can be done using items *First time* and *Time interval.* See paragraph 2.5.3 how to do that. Now, we mention briefly what kind of information the user can get choosing some type of output.

2.5.1 Graphical

If *On* option is specified for graphical output then another window with title "Grid" and items *On,oFf* appears. At this moment the user can select if the grid will be displayed on the screen simultaneously with the trajectory(ies) of the procedure(s). On one hand this option can improve the information available on the other hand less is sometimes more. Having set the Grid option the user can change the c parameter following the editing instructions (see 2.6). The c is the number of screen pixels between two time moments on the time axis that will be left on the screen when the procedure(s) will be displayed. It can be any number less or equal to 50. As the result the path of the selected procedure (or paths of selected procedures in the case of comparisons) is drawn on the screen. Time axis is the horizontal line with the points denoting consecutive time moments. There is the *Time interval* period between each of these two points. On the vertical axis the minimum and maximum values of selected procedures are given. If the whole path can't be situated on the screen then the user may study the rest of it by pressing the key 'y'. If he presses any other key on his keyboard he will finish graphical exhibition of his results.

2.5.2 Numerical and Disk

If *On* option is specified under these items the numerical information will be given. In the *Numerical* case the values of the procedure will be given at twenty consecutive time moments. If the user compare more than one method the results of all the methods chosen under the *Comparison* command (see 2.2.2) will be performed. If he needs to exhibit one method only then more information will be available. There appears six columns on the screen. The first one of them denotes the *Number of iterations* (see 2.3.1). As was mentioned each method is provided from time 0 to Stoptime *Number of iterations* times. Thus we have *Number of iterations* sample paths for each selected method. In the second column there are given the time moments at that the information is given. In the third one the values of the last sample path are performed. In the fourth column there are the averages of the values through all the paths. In the fifth there are given sample variances and finally in the last column we get these sample variances multiplied by square root of time. Information concerning the values in the next time moments is available up to the *Stoptime*. If it is not on the screen the user can continue by pressing 'y' on his keyboard. Setting the Disk output on then all this information will be stored in the disk file together with the information concerning the stored method. The name of the disk file can be changed under the *Output data file* item.

2.5.3 First time and Time interval

Using these two items the user can define the time moments at that he needs his results to be resumed. Choosing *First time* item he can change the first time moment and choosing *Time interval* he can change the interval between two successive moments at that the information will be given. Eediting instructions (see 2.6) can be used in order to change any of these values. For example, if the user needs to study his results only at times 50,60,70,... he would select 50 in the First time item and 10 under the Time interval item.

2.6 Editing instructions

There can appear two alternatives when the user edits. If he edits one item then he enters the value or name that he wants and by pressing Enter on his keyboard this value or name will be accepted. If he edits more then one item he would use Enter to switch between items. Entering the right value in the specified window and pressing Enter this value is accepted. All the values are accepted after pressing Esc key. The user should choose his values or names carefully. Although the program controls some values it could happen that his value will not be recognized as the bad one. This can cause the runtime error of the program.

2.7 How to move in the menu structure

The way is quite simple. The user can use arrow keys to move to the desired item and then press Return. He may directly select the menu command by pressing one of the highlighted letters. Being in some sub menu he can return to the previous level by pressing Esc key.

Acknowledgement: This software would be considerably not so easy to handle without MCRT.TPU "window" unit which was developed by   Martin

May.

References:

[1]Barlow,R.E.,Bartholomew,D.J.,Bremner,J.M.,and   Brunk,H.D.(1972). Statistical Inference Under Order Restrictions.Wiley,London.

[2]Bather,J.A.(1989),Stochastic approximation: A generalization of the Robbins-Monro procedure, Proceedings of the fourth Prague symposium on asymptotic statistics, ed. P.Mandl,M.Huskova, Univ. Publ. House

[3]Blum   J.R.(1954),Multidimensional   stochastic   approximation procedures. Ann. Math. Statist. ,737-744.

[4]Derman, C.(1954), Non-parametric up-and-down experimentation, Ann. Math Statist. 28,795-797.

[5]Dixon,Mood (1948), A method for obtaining and analyzing sensitivity data, J. Ammer. Stat. Assoc.,43,109-126.

[6]Dupač,V.(1981), About stochastic approximation, Kybernetika 17, appendix, 44 pp. (in czech)

[7]Dupač,V.(1984), Stochastic approximation. In Handbook of Statistics, vol. 4,Krishnaiah P.R., Sen P.K. editors, Elsevier Science Publ., Amsterdam, 515-529.

[8]Dupač,V.(1987).Quasi-isotonic   regression   and   stochastic approximation. Metrika,vol 34,page 117-123.

[9]Ermoliev,Yu.,M.(1976), Methods of Stochastic Programming Nauka, Moskva. (in russian)

[10]Ermoliev,Yu.,M.(1988), Stochastic quasigradient methods and their application in systems optimization, IIASA Working Paper-81-2. Numerical Techniques for Stochastic Optimization, Ermoliev,Wets editors, Springer-Verlag,

[11]Gaivoronski, A.(1988), Interactive program SQG-PC for solving stochastic programming problems on IBM PC/XT/AT compatibles-user guide,Numerical Techniques for Stochastic Optimization, Ermoliev,Wets editors, Springer-Verlag,

[12]Huber(1967).The behavior of MLE under non standard conditions. Proceedings of the Fifth Berkeley Symposium on Math. Statistics and Prob. theory. Univ. of California Press 1967.

[13]Hušková, M.(1988),Recursive M-test for detection of change, Sequential Analysis 7(1), 75-90.

[14]Charamza, P.(1984), Stochastic approximation on the lattice. Diploma work on the Fac. of math. and physics Charles Univ., Prague, (in czech)

[15]Charamza, P.(1989), Non-standard stochastic approximation scheme, IIASA Working Paper 1989.

[16]Charamza, P.(1989), Robust isotonic regression, Proceedings of the 6th European Young Statisticians Meeting 1989, ed. M.Hala, M.Maly

[17]Charamza, P.(1989) Isotonic regression and stochastic approximation. PhD thesis on the Faculty of math. and physics, Charles Univ. Prague (in czech)

[18]Kiefer J.,Wolfowitz J. (1952),Stochastic estimation of the maximum of a regression function. Ann. Math. Statist. 23,1952, 462-466

[19]Kirchen, A.(1982),Überlegungen zur eindimensionalen stochastische approximation. Diploma work on University Bonn.

[20]Loeve,M.,Probability Theory (1963),New York,Van Nostrand.

[21]Menendez,Salvador (1987), An algorithm for isotonic median regression, Computational Statistics and Data Analysis 5,399-406.

[22]Mukerjee,H.G.(1981).A stochastic approximation by observation on a discrete lattice using isotonic regression.The Annals of Math.Statistics,vol 9,n.5,page 1020-1025

[23]Nemirovski,A.S.,Polyak,B.T.,Cybakov A.B.(1984). Signal processing using non-parametric maximum likelihood method.Problemy peredaci informaciji,vol XX, (in russian)

[24]Nevelson,Khasminskij,(1972),Stochastic Approximation and recursive estimation,Moscow, Nauka. (in russian)

[25]Pflug, G.(1988),Stepsize rules, stopping times and their implementation in stochastic quasigradient methods. In Numerical Techniques for Stochastic Optimization, Ermoliev,Wets editors, Springer-Verlag,353-372.

[26]Robbins H.,Lai T.L.(1981),Consistency and asymptotic efficiency of slope estimates in stochastic approximation schemes. Z.Wahrsch.Verw.Gebiete 56,N.3,pp 329-360

[27]Robbins H.,Monro S. [1951], A stochastic approximation method. Ann.Math.Statist. 22,1951,400-407.

[28]Schmetterer,L.(1979), From stochastic approximation to the stochastic theory of optimization. Ber. Math.-Statist:Sekt. Forsch. graz, number 27.

[29]Tierney,L.(1983), A space-efficient recursive procedure for estimating a quantile of an unknown distribution, Siam J. Sci. Stat. Comput., vol. 4, No 4.

[30]Wasan M.T. (1969), Stochastic Approximation, Cambridge University Press.

[31]Wichman,P.,Hill,D.(1987), Building a random number generator, Byte 12, March 87, No. 3, 127-128.

LEXICOGRAPHIC DUALITY IN LINEAR
OPTIMIZATION

I.I.Eremin

Institute of Mathematics and Mechanics, Ural
Branch of the USSR Academy of Sciences, 620219
Sverdlovsk, USSR

By lexicographic maximization of the system of functions $\{f_i(x)\}_0^k$ on a set M corresponding to permutation $p = (i_0,...,i_k)$ we mean the problem: Find $\tilde{x} \in M$ such that vector $[f_{i_0}(\tilde{x}),...,f_{i_k}(\tilde{x})]$ is a p-lexicographic maximum on the set

$$Y=\{[y_0,...,y_k] \mid y_t=f_{i_t}(x), \ x\in M, \ t=0,...,k\}.$$

Ordering "$\leq$" in Y is defined as follows: $y \leq z$ if $y_k < z_k$, or $y_k = z_k$ and $y_{k-1} < z_{k-1}$, etc. Inequality $y \leq z$ includes equality $y=z$ corresponding to the case $y_t=z_t$ $\forall t$.

The problem formulated above we denote as

$$\max_{x \in M} F(x), \text{ where } F(x)=[f_{i_0}(x),...,f_{i_k}(x)]^T.$$

1.Consider the following linear problem

$$\max_p \left\{ \begin{bmatrix} (c_0,x) \\ \vdots \\ (c_k,x) \end{bmatrix} \mid Ax \leq b_0 + \sum_{j=1}^{l} r_j b_j, \ x \geq 0 \right\},$$

here $r_j > 0$, $j=1,...,l$, are positive parameters. Let

$$C_0^T = \begin{bmatrix} c_0^T \\ c^T \end{bmatrix}, \quad C^T = \begin{bmatrix} c_1^T \\ \vdots \\ c_k^T \end{bmatrix}, \quad B_0 = [b_0, B], \ B = [b_1,...,b_l].$$ Then the above

problem becomes

$$\mathbb{L}_{p,r}:\max_p \{C_0^T x \mid Ax \leq B_0 r^0 \ (=b_0+Br), \ x \geq 0\}, \tag{1}$$

where $p=(0,i_1,...,i_k)$, $r^0=[1,r_1,...,r_l]^T>0$. We introduce the dual of $\mathbb{L}_{p,r}$ as

$$\mathbb{L}^*_{q,R}:\min_q\{B_0^T u \mid A^T u \geq C_0 R^0 \ (=c_0+CR), \ u\geq 0\},\tag{2}$$

where $q=(0,j_1,...,j_k)$, $R^0=[1,R_1,...,R_k]^T>0$.

Note that permutations p and q in $\mathbb{L}_{p,r}$ and $\mathbb{L}^*_{q,R}$, as well as parameters r and R, are independent so that no regular duality between these problems can exist. A more concrete definition is needed here to construct substantial duality.

The notion of contraction of criteria (as there positive linear combination) is well-known in multicriteria optimization as well as its relation to the Pareto optimization. Using this notion we put $\mathbb{L}_{p,r}$ and $\mathbb{L}^*_{q,R}$ into correspondence with the problems

$$\max\{(c_0,x)+(C\bar{R},x) \mid Ax\leq b_0+Br, \ x\geq 0\},\tag{3}$$

$$\min\{(b_0,u)+(B\bar{r},u) \mid A^T u\geq c_0+CR, \ u\geq 0\}.\tag{4}$$

Objective function in these problems are obtained by contraction of criteria in (1) and (2) using vectors $\bar{R}^T=[\bar{R}_1,...,\bar{R}_k]>0$ and $\bar{r}^T==[\bar{r}_1,...,\bar{r}_l]>0$.

We formulate the following known result (see[1, Theorem 1]).

Lemma 1. If problem (1) is solvable, then

$$\text{Arg } \mathbb{L}_{p,r}=\text{Arg } (3)\tag{5}$$

for some $\bar{R}>0$. Similarly, if problem (2) is solvable then

$$\text{Arg } \mathbb{L}^*_{q,r}=\text{Arg } (4)\tag{6}$$

for some $\bar{r}>0$.

The essence of substantial duality for lexicographic optimization problems (1) and (2) is related to the fact that one can adjust parameters R and $\bar{R}$, as well as r and $\bar{r}$, to ensure equalities (5) and (6) with $r=\bar{r}$ and $R=\bar{R}$.

The explicit formulation of this result will be given later.

Lemma 2. Let problem (1) be consistent. Then it is p-solvable iff

$$c_{i_s} \in \text{cone}\{-c_{i_{s+1}},...,-c_{i_k}; a_1^T,...,a_m^T\} - R_n^+, \quad s=0,...,k. \tag{7}$$

Similar condition for problem (2) is

$$b_{j_t} \in \text{cone}\{-b_{j_{t+1}},...,-b_{j_l}; h_1,...,h_n\} + R_m^+, \quad t=0,...,l. \tag{8}$$

Here

$$\begin{bmatrix} a_{.1} \\ : \\ a_m \end{bmatrix} = [h_1,...,h_n] = A.$$

This result follows immediately from the solvability conditions in linear programming expressed in the form of consistency of both primal and dual problems.

Consider the problems

$$\max\{(c_0,x)+(CR,x) \mid Ax \le b_0 + Br, \ x \ge 0\}, \tag{9}$$

$$\min\{(b_0,u)+(Br,u) \mid A^T u \ge c_0 + CR, \ u \ge 0\}. \tag{9}^*$$

Theorema 1 Let problems

$$\mathbb{L}_{ij}:\max\{(c_i,x) \mid Ax \le b_j, \ x \ge 0\}, \ j=0,...,l, \ i=0,...k$$

be solvable. If the inequality systems

$$Ax \le 0, \ x \ge 0, \ (c_i,x) \ge 1, \ i=1,...,k; \tag{10}$$

$$A^T u \ge 0, \ u \ge 0, \ (b_j,u) \le -1, \ j=1,...,l \tag{11}$$

are consistent, then there exists $R>0$ and $r>0$ such that

$$\text{Arg }(9)=\text{Arg } \mathbb{L}_{p,r} \ne \emptyset,$$

$$\text{Arg }(9)^*=\text{Arg } \mathbb{L}_{q,R}^* \ne \emptyset$$

and, as a consequence, opt $(9)=$opt $(9)^*$.

Proof. This theorem essentially reduces to the possibility of

identifying parameters $\bar{R}$ and R, as well as $\bar{r}$ and r, in problems (3) and (4) to provide inequalities (5) and (6). First, using exact penalty functions method (see [2, Theorem 2]) we will show how to

choose vector $\bar{R}>0$ to provide the validity of Lemma 1 i.e. that of equality (3). Put $M(r)=\{x\geq 0 \mid Ax\leq b_0+Br\}$. Consider the problem

$$\max\{(c_{i_k},x) \mid x\in M(r)\}, \qquad\qquad (i_k)$$

$$\max\{(c_{i_{k-1}},x) \mid x\in \text{Arg }(i_k)\}, \qquad\qquad (i_{k-1})$$

$$\cdots\cdots\cdots\cdots\cdots\cdots\cdots\cdots\cdots\cdots\cdots\cdots\cdots\cdots$$

$$\max\{(c_{i_1},x) \mid x\in \text{Arg }(i_2)\}, \qquad\qquad (i_1)$$

$$\max\{(c_0,x) \mid x\in \text{Arg }(i_1)\}. \qquad\qquad (i_0)$$

Here problem (1) corresponds to problem (i_0). Denote $\alpha_{i_k}=\text{opt}(i_k)$.

Then problem (i_{k-1}) becomes

$$\max\{(c_{i_{k-1}},x) \mid x\in M(r), -(c_{i_k},x)\leq-\alpha_{i_k}\}. \qquad\qquad (12)$$

Let $\bar{u}_k\geq 0$ be the dual estimate of the last inequality in (12).

Then for $R_k^0>\bar{u}_k$ problem (12) will be equivalent (i.e. it will have the same optimal set) to the problem

$$\max\{(c_{i_{k-1}},x)+R_k^0(c_{i_k},x) \mid x\in M(r)\}. \qquad\qquad (13)$$

But we want that R_k^0 be independent of r. When can it be possible? One possible answer can be given immediately: max u_k on the admissible set of the problem dual to (12) must be finite. The problem dual to (12) is as follows:

$$\min\{(b_0+Br,v)-u_k\alpha_{i_k} \mid A^Tu-u_kc_k\geq c_{i_{k-1}}, v\geq 0, u\geq 0\}. \qquad\qquad (14)$$

Now we will find out when the problem

$$\max\{u_k \mid A^Tv-u_kc_{i_k}\geq c_{i_{k-1}}, v\geq 0, u_k\geq 0\} \qquad\qquad (15)$$

is solvable, i.e. $\text{opt}(15)<+\infty$. The last condition holds iff the restrictions of the problem dual to (15) are consistent. These restrictions are of the form

$$A^Tx\leq 0, x\geq 0, (c_{i_k},x)\geq 1. \qquad\qquad (16)$$

Consistency of this system is provided by conditions (10).

Problem (i_{k-2}) can be written in the form

$$\max\{(c_{i_{k-2}},x) \mid x\in M(r), -(c_{i_{k-1}},x)+R_k^0(c_{i_k},x)\leq-\alpha_{k-1}\} \qquad\qquad (17)$$

where $\alpha_{i_{k-1}}$ =opt(13). Here we can similarly derive the condition of choosing $R^0_{i_{k-1}}$ independent of r and such that problem (17), i.e. problem (i_{k-2}), is equivalent to the problem

$$\max\{(c_{i_{k-2}},x)+R^0_{i_{k-1}}[(c_{i_{k-1}},x)+R^0_k(c_{i_k},x)] \mid x \in M(r)\}.$$

We should need the condition of consistency of the system

$$Ax \leq 0, \quad x \geq 0, \quad (c_{i_{k-1}}+R^0_k c_{i_k},x) \geq 1. \tag{18}$$

Moving along the sequence of problems (i_s), $s=k,...,1$, we obtain that problem (i_0), i.e. problem (1), is equivalent to the problem

$$\max\{(c_0+R^0_1[c_{i_1}+R^0_2[c_{i_2}+...+R^0_k c_{i_k}]...],x) \mid x \in M(r)\},$$

where $R^0_1,...,R^0_k$ are independent of r and satisfy the conditions

$$A^T x \leq 0, \quad x \geq 0, \quad (c_{i_k},x) \geq 1;$$

$$A^T x \leq 0, \quad x \geq 0, \quad (c_{i_{k-1}}+R^0_k c_{i_k},x) \geq 1;$$

. . . .

$$A^T x \leq 0, \quad x \geq 0, \quad (c_{i_1}+R^0_2[c_{i_2}+...+R^0_k c_{i_k}]...],x) \geq 1. \tag{19}$$

One can always take $R^0_t \geq 0$ $\forall t$. Then conditions (19) follow from (10). Thus we have shown that one can choose vector $\bar{R}^T=[\bar{R}_1,...,\bar{R}_k]$, where $\bar{R}_1=R^0_1,...,\bar{R}_k=R^0_k$, which is independent of r and ensures the equivalence of problems (1) and (3).

Analogously, if system (11) is consistent one can choose $\bar{r}>0$ independent of R.

Parameters $\bar{R}$ and $\bar{r}$ mentioned above are those realizing the needed concordances $R=\bar{R}$, $r=\bar{r}$. The proof of Theorem 1 is completed.

Thus we can conclude with the following scheme

$$\begin{bmatrix} C^{T} & 0 \\ A & B_{0} \end{bmatrix} \rightarrow \mathbb{L}_{p,r} \xrightarrow{\overset{Arg}{\sim}} \mathbb{L}_{p,r,R}$$

$$(*)\downarrow\uparrow \qquad\qquad (*)\downarrow\uparrow$$

$$\begin{bmatrix} B^{T} & 0 \\ A^{T} & C_{0} \end{bmatrix} \rightarrow \mathbb{L}^{*}_{q,R} \xrightarrow{\overset{Arg}{\sim}} \mathbb{L}_{q,R,r}$$

where $\mathbb{L}_{p,r,R}$ corresponds to (9) with $\bar{R}=R$ and $\mathbb{L}^{*}_{q,R,r}$ corresponds to

(9)* with $\bar{r}=r$. In this scheme extended frames $\begin{bmatrix} C^{T} & 0 \\ A & B_{0} \end{bmatrix}$ and $\begin{bmatrix} B^{T} & 0 \\ A & C_{0} \end{bmatrix}$ of

dual objects are put into correspondence with the substantial

objects of lexicographic optimization $\mathbb{L}_{p,r}$ and $\mathbb{L}^{*}_{q,R}$. The latter

problems are scalarized with the aid of positive vectors R and r

so that their optimal sets are conserved and scalarized problems

are dual to each other.

2. Now we will consider linear program in usual setting but

with the restrictions ordered with respect to their importance

(both in primal and dual problem). These orderings define in a

unique fashion the choice of maximal consistent subsystems (MCS)

of the restriction systems $Ax\leq b$, $x\geq 0$ and $A^{T}u\geq c$, $u\geq 0$ of the

problems

$\quad$ $\mathbb{L}:\max\{(c,x)\mid Ax\leq b, \ x\geq 0\}$,

$\quad$ $\mathbb{L}^{*}:\min\{(b,u)\mid A^{T}u\geq c, \ u\geq 0\}$.

Let these orderings be defined by permutations $(j_{1},...,j_{m})$ and

$(i_{1},...,i_{n})$. It means that inequality in the system $Ax\leq b$ numbered

j_{s+1} is "more important" than inequality numbered j_{s} and,

similarly, inequality in the system $A^{T}u\geq c$ numbered i_{t+1} is "more

important" than inequality numbered i_{t}, $s=1,...,m-1$; $t=1,...,n-1$.

MCS is constructed by adding inequalities beginning with j_{m}-th

inequality (condition $x\geq 0$ is included into MCS automatically). For

the sake of simplicity we assume that MCS's thus constructed are

of the form

$(a_{j_s},x) \le b_{j_s}$, $x \ge 0$, $s=k+1,...,m$;

$(h_{i_t},u) \ge c_{i_t}$, $u \ge 0$, $t=l+1,...,n$.

Left-hand sides $(a_{j_s},x)-b_{j_s}$, $s=1,...,k$, of the remaining inequalities of system $Ax \le b$ (similarly, $(h_{i_t},u) \ge c_{i_t}$, in system

$A^T u \ge c$) are considered to be criteria minimized (resp., maximized) according to the orderings given above (i.e. from the right). These lists of criteria are complemented from the left by functions (c,x) and (b,u), respectively, i.e. these functions are the least important.

Put

$$A_0 = \begin{bmatrix} a_{j_{k+1}} \\ \vdots \\ a_{j_m} \end{bmatrix}, \quad B_0 = [h_{i_{l+1}},...,h_{i_n}], \quad b^0 = \begin{bmatrix} b_{j_{k+1}} \\ \vdots \\ b_{j_m} \end{bmatrix},$$

$$c^0 = \begin{bmatrix} c_{i_{l+1}} \\ \vdots \\ c_{i_n} \end{bmatrix}, \quad C^T = \begin{bmatrix} -a_{j_1}^T \\ \vdots \\ -a_{j_k}^T \end{bmatrix}, \quad B=[b,h_{i_1},...,h_{i_l}].$$

According to the partition of matrix A into horizontal and vertical submatrices we have

$$A = \begin{bmatrix} A_0 \\ A_1 \end{bmatrix} = [B_0 \, B_1].$$

This partition induces the partitions of vectors: $x = \begin{bmatrix} x^0 \\ x^1 \end{bmatrix}$, $u = \begin{bmatrix} u^0 \\ u^1 \end{bmatrix}$.

We put $p=(0,j_1,...,j_k)$, $q=(0,i_1,...,i_l)$ and formulate the following lexicographic optimization problems:

$\mathbb{L}_{p,r} : \max_p \{ C^T x \mid A_0 x \le b^0, \ x \ge 0, \ x^1 \le r \}$,

$\mathbb{L}_{q,R}^* : \min_q \{ B^T u \mid B_0^T u \ge c^0, \ u \ge 0, \ u^1 \le R \}$.

The analogs of problems (9) and (9)* for these problems are of the form

$$\mathbb{L}_{p,r,R} : \max_p \{ (c,x)-(A_1 R,x) \mid A_0 x \le b^0, \ x \ge 0, \ x^1 \le r \} \qquad (20)$$

$$\mathbb{L}_{q,R,r}^\# : \min_q \{ (b,u)-(B_1 r,u) \mid B_0^T u \ge c^0, \ u \ge 0, \ u^1 \le R \}. \qquad (21)$$

Put $J_0 = \{ j_{k+1},...,j_m \}$, $J_1 = \{ j_1,...,j_k \}$, $I_0 = \{ i_{l+1},...,i_n \}$, $I_1 =$

$=\{i_1,...,i_l\}$. Let e_i and e_j^* be basis unit vectors in R^n and R^m, respectively. Consider the requirements

$$c\in \text{cone }\{\{a_j\}, \, j\in J_0; \, \{e_i\}, \, i\in J_1\}-R_+^n, \qquad (22)$$

$$b\in \text{cone }\{\{h_i\}, \, i\in I_0; \, \{e_j^*\}, \, j\in I_1\}+R_+^m \qquad (23)$$

and the systems

$$B_0 x \le \begin{bmatrix} 0 \\ -E_k \end{bmatrix}, \, x^0 \ge 0; \qquad (24)$$

$$A_0^T u \ge \begin{bmatrix} 0 \\ E_l \end{bmatrix}, \, u^0 \ge 0, \qquad (25)$$

where $E_k=[1,...,1]^T$, $\underset{\langle--k--\rangle}{}$ $E_l=[1,...,1]^T$. $\underset{\langle--l--\rangle}{}$

Theorem 2. Let (22) and (23) hold and let system (24) and (25) be consistent. Then

$$\text{Arg } L_{p,q}=\text{Arg}(20)\neq\emptyset, \quad \text{Arg } L_{q,R}^*=\text{Arg}(21)\neq\emptyset$$

for some $R>0$, $r>0$, moreover,

$$\text{opt}(20)+(b^1,R)=\text{opt}(21)+(c^1 r),$$

where $c=\begin{bmatrix} c_0^0 \\ c_1 \end{bmatrix}$, $b=\begin{bmatrix} b_0^0 \\ b_1 \end{bmatrix}$ according to the partition of A into A_0, A_1 and B_0, B_1.

The proof of this theorem is analogous to that of Theorem 1.

We conclude section 2 with the following scheme:

$$L \;\to\; L_{p,r} \;\overset{\underset{\sim}{\text{Arg}}}{\to}\; L_{p,r,R}$$

$$(*)\downarrow\uparrow \qquad\qquad (\#)\downarrow\uparrow$$

$$L^* \;\to\; L_{q,R}^* \;\overset{\underset{\sim}{\text{Arg}}}{\to}\; L_{q,R,r}^{\#}$$

References

1. Eremin I.I. On the problems of successive programming // Sib. matem. jurn. 1973. T.14, No.1. P.53-63. (in Russian).
2. Eremin I.I. Penalty function method in convex programming // Dokl. AN SSSR. 1967. T.173, No.4. P.748-751. (in Russian).

DUAL OPTIMIZATION OF DYNAMIC SYSTEMS

R. Gabasov
Byelorussian State University
220080 Minsk, USSR

F. Kirillova
Institute of Mathematics
220604 Minsk, USSR

Abstract. Three pairs of adjoint control and observation problems and a pair of adjoint control and identification problems are investigated. On the base of constructive theory of extremal problems created by the authors and their collaborators a method of constructing guaranteeing program optimal solutions using observation and control procedures is set.

1. Introduction.

Dual problems as regards Kalman controllability and observability problems of dynamic systems are well known in the qualitative theory of optimal processes [1-4]. In the paper another point of view on connection between control, observation and identification problems is presented.

Classical results of the theory of optimal processes were obtained for determined models. It allowed to solve control and observation problems separately [1,2,5]. The first incompletely defined models of optimization processes appeared in frames of stochastic optimal control [6,7]. A.A. Feldbaum [8] paid attention to the dual role of control in such problems . Problems in which observation and control procedures are also can be separated in the conditions of uncertainty were selected in [9] . Some uncertain models aimed at obtaining guaranteed results came to the control theory from the game theory [4,10,12]. The accent here is made on structures of sets of possible values of indefinite elements. Two mentioned types of models are consonant to two classical models of the theory of extremal problems. On the base of constructive theory of extremal problems [13,14] a method of construction of guaranteeing program optimal solutions using observation and control procedures is set in the paper.

2. Adjoint Problems of Control, Observability and Identification.

1. Let the set

$$X_o = \{x \in \mathbb{R}^n: Dx = b, \; d_* \leq x \leq d^*\}, \quad \{d_*, d^* \in \mathbb{R}^n; \; b \in \mathbb{R}^m; \; D \in \mathbb{R}^{n \times m}\},$$

a piecewise-continuous function $y(t)$, $t \in T = [0, t^*]$; an $n \times n$-matrix A, vectors $c, h \in \mathbb{R}^n$, scalars g_*, g^* are given.

Consider the problem of optimal choice of initial state of dynamic system in which the motion $y(t)$, $t \in T$, is realized, i.e.

$$c^T z \longrightarrow max, \tag{1}$$

$$\dot{x} = Ax, \; x(0) = z, \tag{2}$$

$$z \in X_o, \tag{3}$$

$$g_* \leq y(t) - h^T x(t) \leq g^*, \; t \in T. \tag{4}$$

Here superscript ^{T} denotes transposition. The adjoint observation problem for (1)-(4) is formulated in the following way.

A priori distribution X_o of initial states of dynamic system (2) is known. Suppose we have the measure device

$$y = h^T x + \zeta, \tag{5}$$

that measures the output signals $h^T x(t)$, $t \in T$, with some error $\zeta(t)$, $t \in T$. The error realizations $\zeta(t)$, $t \in T$, are supposed to be piecewise-continuous functions satisfying the conditions

$$g_* \leq \zeta(t) \leq g^*, \; t \in T.$$

The set X^* of initial states $z \in X_o$ which together with some errors $\zeta(t)$, $t \in T$, can generate the observed signal $y(t)$, $t \in T$, is called a posteriori distribution of initial states of system (2).

The simplest number characteristic of the set X^* is the extention $max \; c^T z, \; z \in X^*$, in a direction c.

The observation problem

$$c^T z \longrightarrow max, \; z \in X^*,$$

will be called an adjoint problem with respect to the optimal problem of the initial states (1)-(4). The both problems are reduced to the semiinfinite extremal problem:

$$c^T x \longrightarrow max, \; b_*(t) \leq a^T(t)x \leq b^*(t), \; t \in T;$$

$$Ax = b, \; d_* \leq x \leq d^*. \tag{7}$$

with finite number of variables and infinite number of constraints.

2. Let in addition a piecewise-continuous function $c(t)$, $t \in T$; n-vectors b, x_o and scalar y^* are given.

Consider the optimal control problem

$$\int_0^{t^*} c(t)u(t)dt \;\longrightarrow\; max, \quad \dot{x} = Ax + bu, \quad x(0) = x_o,$$

$$g_* \le y - h^T x(t^*) \le g^*, \quad |u(t)| \le 1, \quad t \in T,$$

(8)

in the class of piecewise-continuous functions $u(t)$, $t \in T$.

Construct the adjoint observation problem with respect to (8). Let a priori distribution Ω_o of piecewise-continuous perturbations $w(t)$, $t \in T$, is known. Thay act on an input of the dynamic system:

$$\dot{x} = Ax + bw(t), \quad x(0) = x_o.$$

(9)

The measure divice (5) indicates signal $h^T x(t^*)$ at the moment t^* with the error ζ that can have a meaning satisfying the conditions

$$g_* \le \zeta \le g^*.$$

(10)

Denote a posteriori distribution of perturbations by Ω^*. It consists of those perturbations $w(t)$, $t \in T$, and only of those that can generate the observed signal y^* with errors ζ_j. The simplest number characteristic of the set Ω^* is the maximum meaning of moments

$$\int_0^{t^*} c(t)w(t)dt$$

of functions of Ω^*.

The calculation of this characteristic reduces to the following extremal problem

$$\int_0^{t^*} c(t)w(t)dt \;\longrightarrow\; \max_w, \quad w(\cdot) \in \Omega^*.$$

(11)

Problems (8),(11) are particular cases of semiinfinite extremal problem

$$\int_0^{t^*} c(t)u(t)dt \;\longrightarrow\; max, \quad b_* \le \int_0^{t^*} a(t)u(t)dt \le b^*,$$

$$|u(t)| \le 1, \quad t \in T,$$

(12)

with infinite number of variables and finite number of general constraints.

3. A terminal problem of constructing optimal control realizing the given motion $y(t)$, $t \in T$, is as follows

$$c^T x(t^*) \;\longrightarrow\; max, \quad \dot{x} = Ax + bu, \quad x(0) = x_o,$$

$$g_* \le y(t) - h^T x(t) \le g^*, \quad |u(t)| \le 1, \quad t \in T.$$

This problem is associated with the following adjoint observation problem

$$c^T x \longrightarrow max, \quad x \in X^*(t^*),$$

intended for calculating the extension of a posteriori distribution $X^*(t^*)$ of the terminal states $x(t^*)$ of system (9) under assumption that the system is affected by piecewise-linear perturbations $w(t)$, $t \in T$, $|w(t)| \leq 1$, $t \in T$, and the signal $y(t)$, $t \in T$, is obtained with errors satisfying inequalities (6).

Mentioned problems of control and observation are particular cases of the infinite extremal problem

$$\int_0^{t^*} c(t)u(t)dt \longrightarrow max, \quad b_*(t) \leq \int_0^{t^*} a(t,\tau)u(\tau)d\tau \leq b^*(t), \tag{13}$$

$$|u(t)| \leq 1, \quad t \in T.$$

Finite algorithms for solving problems (7),(12),(13) and their application for solving various problems of optimal control are presented in [14]. According to results of items 1-3 the algorithms can be used for solving observation problems too.

4. Let $n \times n$ matrices A_0, $A_1, \ldots,$ A_q; n-vectors b_0, $b_1, \ldots,$ b_q; piecewise-continuous functions $u_0(t)$, $u_1(t), \ldots,$ $u_q(t)$, $t \in T$, and a set $W = \{w \in R^q: Gw = f, w_* \leq w \leq w^*\}$, $(f \in R^l)$, of the values of parameter $w = (w_1, \ldots,$ $w_q)$, on which the matrix A, the vector b and the control $u(t)$, $t \in T$, of the dynamic system

$$\dot{x} = Ax + bu, \quad x(0) = x_0,$$

$$A = A_0 + \sum_{i=1}^q w_i A_i, \quad b = b_0 + \sum_{i=1}^q w_i b_i, \tag{14}$$

$$u(t) = u_0(t) + \sum_{i=1}^q w_i u_i(t), \quad t \in T,$$

depend, be given.

The problem of optimal control of the dynamic system parameters is as follows

$$c^T w \longrightarrow max,$$

$$\dot{x} = (A_0 + \sum_{i=1}^q w_i A_i)x + (b_0 + \sum_{i=1}^q w_i b_i)(u_0(t) + \sum_{i=1}^q w_i u_i(t)), \tag{15}$$

$$x(0) = x_0, \quad t \in T; \quad g_* \leq y - h^T x(t^*) \leq g^*, \quad w \in W.$$

Let us formulate the problem of identification adjoint with respect to problem (15).

Dynamic system (14) with an unknown value of vector w is functioning at the segment T. A priori distribution W of the dynamic system parameters is known. To make this distribution more correct at the moment t^* the signal $y^*=y(t^*)$ is written by the measuring device (5). A posteriori distribution W^* consisting of only those parameters $w \in W$, which together with the mesurement error $\zeta=\zeta(t^*)$ satisfying inequalities (10) can generate an output y^* of the measuring device (5) corresponds to y^*. Computation of the extension of the set W^* along a direction c:

$$c^T w \longrightarrow max, \quad w \in W^*, \tag{16}$$

is called the problem of identification of the dynamic system. It is adjoint with respect to the control problem (15).

The problems (15),(16) represent particular cases of nonlinear programming problem

$$c^T x \longrightarrow max, \quad f(x) = 0, \quad d_* \leq x \leq d^*. \tag{17}$$

Special methods for solving problem (17) based on network models of nonlinear functions are presented in [16].

To make the structures of a posteriori distributions X^*, $X^*(t^*)$, Ω^*, W^* more correct besides several directions one can introduce quantization of these sets in various ways.

Above four special problems of control, observation and identification were chosen only for simplicity and definiteness. Their various generalizations are evident. It is important to emphasize that according to the described approach all these problems are primal. It is known that primal form for control problems and dual form for observation problems are widely used. From the point of view of the costructive theory of extremal problems primal forms are more natural. Dual forms play an important role but only as aixiliary problems.

2. Dual Optimization of Dynamic Systems.

1. Let be given: sets $\check{X}_o \subset R^n$, $X^* = \bigcap\limits_{i=1}^m X_i^*$, $X_i^*=\{x \in R^n: h_i(x) \geq 0\}$; families $\check{\Omega}(\cdot)$, $U(\cdot)$, $\Xi(\cdot)$ of p-vector-functions $\omega(\cdot)=(\omega(t), t \in T=[t_*, t^*])$, r-vector-functions $u(\cdot)=(u(t), t \in T)$ and k-vector-functions $\zeta(\cdot)=(\zeta(t), t \in T)$ respectively. The sets $\check{X}_o$, $\check{\Omega}(\cdot)$ will be called a priori

distributions of initial states and perturbations of the dynamic system

$$\dot{x} = f(x, u, \omega, t), \quad x(t_*) \in \check{X}_o, \quad \omega(\cdot) \in \check{\Omega}(\cdot) \quad (x \in R^n, u \in R^r). \tag{1}$$

Assume that the only trajectory $x(t) = x(t \,|\, x_o, u(\cdot), \omega(\cdot))$, $t \in T$, of system (1) corresponds to each control $u(\cdot) \in U(\cdot)$, the initial state $x(t_*) = x_o \in \check{X}_o$ and the perturbation $\omega(\cdot) \in \check{\Omega}(\cdot)$. The totality $\check{X}(t) = \check{X}(t \,|\, u(\cdot)) = \{x(t \,|\, x_o, u(\cdot), \omega(\cdot)), \; x_o \in \check{X}_o, \; \omega(\cdot) \in \check{\Omega}(\cdot)\}$, $t \in T$, describes the law of variation in time of a priori distribution of states of system (1).

A control $\check{u}(\cdot)$ is a priori-admissible if

$$\check{X}(t^* \,|\, \check{u}(\cdot)) \subset X^*. \tag{2}$$

The quality of a priori-admissible control will be estimated with respect to the value of functional

$$J(\check{u}) = \min h_o(z), \quad z \in \check{X}(t^*). \tag{3}$$

A priori-optimal control $\check{u}^o(\cdot)$ will be determined by the equality

$$J(\check{u}^o) = \max_{\check{u}(\cdot)} J(\check{u}). \tag{4}$$

2. To raise the effeciency (4) of control of the dynamic system in conditions of uncertainty let us introduce a measuring device

$$y = c(x, u, \zeta) \quad (y \in R^l). \tag{5}$$

With reference to realizations of errors of the measuring $\zeta(t)$, $t \in T$, we shall assume that any element of the family $\Xi(\cdot)$ can be found on this place.

Let the initial state x_o, the perturbation $\omega(\cdot)$, the functions of errors $\zeta(\cdot)$ be realized after choice of the control $u^*(\cdot) \in U(\cdot)$ and the measuring device wrote the signal

$$y^*(\cdot) = (y^*(t), \; t \in T_\theta = [t_*, \theta]), \quad \theta \leq t^*.$$

The totality $\hat{X}_o = \hat{X}_o(u^*)$, $\hat{\Omega}(\cdot) = \hat{\Omega}(\cdot \,|\, u^*)$ of all elements $x_o \in \check{X}_o$, $\omega(\cdot) \in \check{\Omega}(\cdot)$, which together with some functions $\zeta(\cdot) \in \Xi(\cdot)$ are able to generate the observed signal $y^*(\cdot)$ will be called a posteriori distribution of initial states and perturbations. Denote by $\hat{X}(t^* \,|\, u^*(\cdot)) = \{x(t^* \,|\, x_o, u^*(\cdot), \omega(\cdot)), \; x_o \in \hat{X}_o, \; \omega(\cdot) \in \hat{\Omega}(\cdot)\}$ a posteriori distribution of terminal state of system (1). Keeping in mind in the sequel the obtaining of the guaranteed result for problem (1)–(5) we calculate the following estimates of the set $\hat{X}(t^*|u^*(\cdot))$:

$$\hat{\alpha}_{\iota} = \hat{\alpha}_{\iota}(u^*) = \min h_{\iota}(z), \quad z \in \hat{X}(t^*|u^*(\cdot)), \quad \iota = \overline{0, m}. \tag{6}$$

The calculation of the estimates (6) will be called problems of observation accompaning the optimization problem (1)–(4) in conditions of uncertainty for chosen $u^*(\cdot)$. In detailed writing such observation problems are the extremal problems

$$\hat{\alpha}_{\iota} = \min_{z, \omega(\cdot), \zeta(\cdot)} h_{\iota}(x(t^*)), \quad \dot{x} = f(x, u^*, \omega, t),$$

$$x(t_*) = z, \quad y^*(t) = c(x(t), u^*(t), \zeta(t), t), \quad t \in T_{\theta}, \tag{7}$$

$$\omega(\cdot) \in \check{\Omega}(\cdot), \quad \zeta(\cdot) \in \Xi(\cdot), \quad z \in \check{X}_0; \quad \iota = \overline{0, m}.$$

Problem (7) for a fixed ι can be treated as a determined problem of optimal control: let be given an input signal $u^*(\cdot)$ and an output signal $y^*(\cdot)$, it is required to find optimal values of initial state $z \in \check{X}_0$, control $\omega(\cdot) \in \check{\Omega}(\cdot)$ and correcting function $\zeta(\cdot) \in \Xi(\cdot)$ when an exact observation over $y^*(\cdot)$ is carried out and a minimal value of the terminal cost functional $h_{\iota}(x(t^*))$ is achieved.

3. Supplement problem (1)–(4) by the measuring device (5). Controls $u^*(\cdot)$ limited by the restrictions $u^*(\cdot) \in U(\cdot)$ are a posteriori admissible if $\alpha_{\iota}(u^*) \geq 0$, $\iota = \overline{1, m}$. A posteriori admissible control $\hat{u}^o(\cdot)$ is called a posteriori-optimal if

$$\hat{\alpha}_0(\hat{u}^o) = \max \hat{\alpha}_0(\hat{u}^*). \tag{8}$$

Problem (1),(2),(8) will be called a determined problem of optimal control accompanying (1)–(5).

It is seen from (1)–(5) that the control in the optimization problem (1)–(4) with the measuring device (5) fulfils two-fold role. On the one hand it decreases uncertainty of the problem, on the other hand it increases the value of the cost functional. Just by this reason A. A. Feldbaum had called such control dual.

4. Consider a special case of problem (1)–(4) when $\check{X}_0 = \{x \in R^n: Gx = f, d_* \leq x \leq d^*\}$, $\check{\Omega}(\cdot) = \emptyset$, $f(x, u, \omega, t) = A(t)x + B(t)u$; $A(t), B(t)$, $t \in T$, are piecewise-continuous $n \times n$ and $n \times r$ matrix functions respectively, $U(\cdot) = \{u(t) \in R^r: \mathcal{L}u(t) = l, u_* \leq u(t) \leq u^*, t \in T\}$, $X_{\iota}^* = \{x \in R^n: h_{\iota}^T x \geq g_{\iota}, \iota = \overline{1, m}\}$, $h_0(x) = h_0^T x$, $c(x, u, \zeta) = Cx + \zeta$, $k = l$, $\Xi(\cdot) = \{\zeta(t) \in R^k: \zeta_* \leq \zeta(t) \leq \zeta^*, t \in T\}$, $\theta = t^*$.

A priori-optimal control $\check{u}^o(t)$, $t \in T$, is the solution of the determined problem

$$J_o(u) = h_o^T x(t^*) \rightarrow max, \quad \dot{x} = A(t)x + B(t)u, \quad x(t_*) = 0,$$

$$h_i^T x(t^*) \geq \check{g}_i, \quad i=\overline{1,m}; \quad u_* \leq u(t) \leq u^*, \quad \mathcal{L}u(t) = l, \quad t \in T,$$

where $\check{g}_i = g_i - \check{\gamma}_i$, $\check{\gamma}_i$ are solution of the linear programming problem

$$\check{\gamma}_i = min \; p_i^T(t^*)z, \quad Gz = f, \quad d_* \leq z \leq d^*, \quad i=\overline{0,m}; \tag{9}$$

$$(p_i^T(t) = h_i^T F(t), \quad dF(t)/dt = AF, \quad F(t_*) = E).$$

Finite methods for solving problem (9) are given in [15]. The value of cost functional on $\check{u}^o(\cdot)$ is equal to

$$J(\check{u}^o) = \check{\gamma}^o + J_o(\check{u}^o).$$

Let $y^*(t)$, $t \in T$, be an observed signal of the measuring device (5) caused by the trajectory $x(t|x_o, u^*(\cdot))$, $t \in T$, and an unknown function of errors $\zeta(t)$, $t \in T$; $z(t) = y^*(t) - \int_{t_*}^{t} CF(t)F^{-1}(\tau)b(\tau)u(\tau)dt$, $t \in T$. Calculate the estimates $\hat{\alpha}_i = min \; h_i^T x(t^*|x_o, u^*(\cdot))$, $i=\overline{0,m}$, $x_o \in \hat{X}_o$.

A posteriori-optimal control $\hat{u}^o(t)$, $t \in T$, is the solution of a determined problem accompanying (1)-(4) with the measuring device (5):

$$J_o(u) = h_o^T x(t^*) \rightarrow max, \quad \dot{x} = A(t)x + b(t)u, \quad x(t_*) = 0,$$

$$h_i^T x(t^*) \geq \hat{g}_i, \quad i=\overline{1,m}; \quad u_* \leq u(t) \leq u^*, \quad \mathcal{L}u(t) = l, \quad t \in T,$$

where $\hat{g}_i = g_i - \hat{\gamma}_i$, $i=\overline{1,m}$; $\hat{\gamma}_i$ is the value of the problem

$$\hat{\gamma}_i = min \; p_i^T(t^*)z, \quad Gz = f, \quad d_* \leq z \leq d^*,$$

$$\zeta_* \leq z(t) - CF(t)z \leq \zeta^*, \quad t \in T; \quad i=\overline{0,m}. \tag{10}$$

A method for solution of problem (10) is given in [15]. We have $J(\hat{u}^o) = J_o(\hat{u}^o) + \hat{\gamma}^o$. The value $J(\hat{u}^o) - J(\check{u}^o)$ characterizes the increase of control efficiency using the measuring device (5).

The second special case of problem (1)-(4) is of interest for applications: $\check{X}_o = \{x_o\}$, $\Omega(\cdot) = \{\omega(t) \in R: \quad \omega(t) = \omega_o(t) + \sum_{i=1}^{q} \omega_i(t)w_i, \quad t \in T\}$,

$f(x,u,\omega,t)=A(t)x+B(t)u+D(t)\omega$, the rest elements are the same as in the item 2.

A priori-optimal control $\breve{u}^0(t)$, $t\in T$, is the solution of the determined problem

$$J_o(u) = h_o^T x(t^*) \longrightarrow max,$$

$$\dot{x} = A(t)x + b(t)u + D(t)\omega_o(t), \qquad x(t_*) = x_o,$$

$$h_i^T x(t^*) \geq \breve{g}_i, \qquad i=\overline{1,m}; \qquad u_* \leq u(t) \leq u^*, \qquad \mathcal{L}u(t) = l, \qquad t\in T.$$

Here $\breve{g}_i = g_i - \breve{\gamma}_i$, $\breve{\gamma}_i$ is the value of the problem

$$\breve{\gamma}_i = min\ a_i^T w, \qquad Gw = f, \qquad d_* \leq w \leq d^*, \qquad i=\overline{0,m}. \tag{11}$$

$$(a_i = (a_{ij},\ j=\overline{1,q}), \qquad a_{ij} = h_i^T \int_{t_*}^{t^*} F(t^*)F^{-1}(t)D(t)\omega_j(t)dt).$$

he value of cost functional on $\breve{u}^0(t)$ is equal to $J(\breve{u}^0)=J_o(\breve{u}^0)+\breve{\gamma}_o$. The finite methods for solving problem (11) are given in [14].

Let $u^*(t)$, $t\in T$, be some admissible control, $x(t|x_o,u^*(\cdot),w)$, $t\in T$, be a trajectory corresponding to the control $u^*(\cdot)$, the initial state x_o and the value of parameter $w\in\breve{W}$; $y^*(t), t\in T$, be an observed signal of the measuring device (5), caused by $x(t|x_o,u^*(\cdot),w)$, $t\in T$, and an unknown function of errors of measuring $\xi(t)$, $t\in T$. Assume

$$z(t)=y^*(t) - \int_{t_*}^{t}CF(t)F^{-1}(\tau)bu^*(\tau)d\tau - CF(t)x_o - \int_{t_*}^{t}CF(t)F^{-1}(\tau)D(\tau)\omega_o(\tau)d\tau.$$

A posteriori-optimal control $\hat{u}^0(t)$, $t\in T$, is the solution of the problem

$$J_o(u) = h_o^T x(t^*) \longrightarrow max,$$

$$\dot{x} = A(t)x + b(t)u + D(t)\omega_o(t), \qquad x(t_*) = x_o,$$

$$h_i^T x(t^*) \geq \hat{g}_i, \qquad i=\overline{1,m}; \qquad u_* \leq u(t) \leq u^*, \qquad \mathcal{L}u(t) = l, \qquad t\in T,$$

where $\hat{g}_i = g_i - \hat{\gamma}_i$, $\hat{\gamma}_i$ is the value of the problem

$$\hat{\gamma}_i = min\ a_i^T w, \qquad Gw = f, \qquad d_* \leq w \leq d^*,$$

$$\xi_* \leq z(t)-Cw^T d \leq \xi^*, \quad t\in T,$$

$$(d=(d_j,\ j=\overline{1,q}), \qquad d_j = \int_{t_*}^{t^*} F(t^*)F^{-1}(t)D(t)\omega_j(t)dt).$$

The value of cost functional on a posteriori-optimal control $\hat{u}^{\circ}(t)$, $t \in T$, is equal to $J(\hat{u}^{\circ}) = J_{\circ}(\hat{u}^{\circ}) + \hat{\gamma}_{\circ}$. The value $J(\hat{u}^{\circ}) - J(\check{u}^{\circ})$ characterizes the increase of control efficiency at the expense of using the measuring device (5).

References

1. Kalman R. On General Theory of Control Systems, Proceedings of the 1th IFAC Congress, vol. 2, Moscow, Academy of Sc., 1961, p. 521-547 (in Russian).

2. Krasovskii N.N. The Theory of Controlled Motion, Moscow, Nauka, 1968 (in Russian).

3. Gabasov R., Kirillova F.M. Qualitative Theory of Optimal Processes. Moscow, Nauka, 1971 (in Russian).

4. Kurzanskii A.B. Control and Observation under Conditions of Uncertainty. Moscow, Nauka, 1977 (in Russian).

5. Pontryagin L.S., Boltyanskii V.G., Gamkrelidze R.V., Mishchenko E.F. Mathematical Theory of Optimal Control. Moscow, Physmatgiz, 1961 (in Russian).

6. Lenning J.H., Battin R.G. Random processes in Problems of Automatic Control. Moscow, NIL, 1958 (in Russian).

7. Pugachev V.S. Theory of Random Functions and its Application in Automatic Control Problems. Moscow, Gostechizdat, 1957 (in Russian).

8. Feldbaum A.A. Fundamentals of Theory of Optimal Automatic Systems. Moscow, GIFML, 1963 (in Russian).

9. Simon H.A. Econometrica, 24, 74(1956).

10. Schweppe F.C. Uncertain Dynamic System. Englewood Cliffs: Prentice Hall, 1973.

11. Krasovskii N.N. Control by Dynamic System. Moscow, Nauka, 1977 (in Russian).

12. Chernousko F.L. Estimating of Phase State of Dynamic Systems. Moscow, Nauka, 1988 (in Russian).

13. Gabasov R., Kirillova F.M. Linear Programming Methods. Parts 1-3. Minsk, BGU Press, 1977,1978,1980 (in Russian).

14. Gabasov R., Kirillova F.M., (Tyatyushkin A.I., Kostyukova O.I., Raketskii V.M.) Constructive Methods of Optimization. Parts 1-4. Minsk, University Press, 1984,1986,1988 (in Russian).

15. Gabasov R., Kirillova F.M., Kostyukova O.I., Pokatayev A.V. in: Constructive Theory of Extremal Problems. Minsk, University Press, 1984 (in Russian).

STOCHASTIC APPROXIMATION VIA AVERAGING:
THE POLYAK'S APPROACH REVISITED

G. Yin*

Department of Mathematics
Wayne State University, Detroit, MI 48202

Abstract. Recursive stochastic optimization algorithms are considered in this work. A class of multistage procedures is developed. We analyze essentially the same kind of procedures as proposed in Polyak's recent work. A quite different approach is taken and correlated noise processes are dealt with. In lieu of evaluating the second moments, the methods of weak convergence are employed and the asymptotic properties are obtained by examining a suitably scaled sequence. Under rather weak conditions, we show that the algorithm via averaging is an efficient approach in that it provides us with the optimal convergence speed. In addition, no prewhitening filters are needed.

Keywords. recursive estimation, averaging, stochastic approximation, weak convergence.

1. Introduction

Due to a wide range of applications in stochastic optimization, there has been growing and renewed interest in studying recursive algorithms of stochastic approximation type (cf. [1-3] and the references therein).

One of the crucial issues is to improve the performance of the recursive procedures and design more efficient algorithms. Recently, much efforts have been devoted to developing algorithms that can be executed by using multiprocessors and parallel processing methods. Asynchronous and distributed schemes, which use convexification extensively, were introduced in Tsitsiklis, Bertsekas and Athans [4-5]. Further analysis on these algorithms, including convergence, rates of convergence, projection algorithms as well as communication through noisy links, was furnished in Kushner and Yin [6]. Another class of algorithms that emphasizes parallel processing aspects was proposed in Kushner and Yin [7], and asymptotic properties were developed by utilizing the weak convergence methods and martingale averaging techniques. A parallel Robbins-Monro type of procedure was studied in Yin and Zhu [8], and the convergence was obtained via a random projection procedure. An extensive survey of the aforementioned topics can be found in [9] and the references therein.

Along another line, various adaptive stochastic approximation algorithms have been proposed and analyzed. Much progress has been made in improving the asymptotic performance of the underlying algorithms.

In a recent paper [10], Polyak utilized the idea of averaging and developed a multi-step procedure. Using the idea of averaging, a scalar-valued adaptive RM algorithm was also developed independently in the work of Ruppert [11]. Unlike most of the existing adaptive procedures, in

---

* This research was supported in part by the National Science Foundation under grant DMS-9022139.

which matrix-valued estimates are needed for the gradient of the underlying function, the algorithm proposed in [10] is rather simple. In comparison with the traditional stochastic approximation algorithms, virtually, no extra computation burdens are created in the new algorithms. It provides us with an easily implementible and efficient procedure.

In spite of the interesting and novel approach, one of the main assumptions used in [10] is that the observation noise processes are uncorrelated. Nevertheless, in various stochastic optimization problems arisen in physical sciences, engineering, economics and social sciences, more often than not, the noise processes are correlated. In such cases, analytical results obtained in [10] are not directly applicable. In fact, it was noted in [10], if the random process is of moving average (MA) type, then some kind of prewhitening technique should be employed. However, this creates further computation complexity.

In this work, our objective is to extend the results of Polyak to include correlated random sequences. We shall show that under much weaker conditions (to be more specific, assuming the noise to be φ-mixing type [12]), the desired results still hold without using prewhitening procedures. The approach we are using is quite different from that of [10]. In lieu of evaluating the second moments, we prove the weak convergence of a suitably scaled sequence and hence derive the asymptotic limit theorem.

The rest of the paper is organized as follows. Polyak's algorithm is given next. Our assumptions and the main theorem will also be stated. Section 3 is devoted to the proof of the results. In Section 4, a couple of simple examples are provided. Finally, some possible extensions and further remarks are made in Section 5.

To proceed, a few words about the notations are in order. In the sequel, K will be used to denote a generic positive constant. Its value may change under different usages. '$\prime$' will be used to denote the transpose of a matrix.

2. Polyak's algorithm

For simplicity, we shall consider the Robbins-Monro type of algorithm only. Let $f : \mathbf{R}^r \mapsto \mathbf{R}^r$, and $\{\xi_n\}$ be a sequence of $\mathbf{R}^r$-valued random variables. Let the approximating sequence $\{x_n\}$ be generated by:

$$x_{n+1} = x_n + a_n(f(x_n) + \xi_n), \tag{2.1}$$

where $\{a_n\}$ is a sequence of positive real numbers known as gain or step size.

Much of the work in adaptive stochastic approximation concerns about the problem of choosing a_n so as to improve the asymptotic performance. Such performance is often measured by the asymptotic covariance matrix S. Some of the treatments for adaptive algorithms with correlated noise can be found in the work of Benveniste, Metivier and Priouret [13] and the references therein.

Actually, the seeking of asymptotic optimal performance can be traced back to the work of Chung [14]. He studied a one dimensional version of the algorithm and showed that if $\Gamma = -1/f_x(\theta)$, then the optimality is achieved. Venter [15] went a step further and designed an adaptive estimation procedure. In addition to estimating θ, a sequence of consistent estimates of $f_x(\theta)$ was also constructed. Lai and Robbins [16] took a some what different approach to attack the problem. They treated the underlying problem as both estimation and control and suggested using a least squares method to estimate $f_x(\theta)$. The above papers all dealt with one dimensional problems, whereas [13, 17] treated multidimensional systems.

Consider the following algorithm:

$$x_{n+1} = x_n + \frac{\Gamma}{n}(f(x_n) + \xi_n).$$

To obtain the asymptotic normality for the correlated noise processes of mixing type (the precise conditions are given in the sequel), we can use the approach in [18]. Let $H = f_x(\theta)$ and define:

$$C_{jk} = \begin{cases} \prod_{l=k+1}^{j}(I + \frac{\Gamma H}{l}), & j \geq k+1; \\ I, & j = k, \end{cases}$$

$$M_n(t) = \frac{[nt]}{\sqrt{n}} \sum_{k=1}^{[nt]} \frac{1}{k} C_{[nt]k} \Gamma \xi_k, \quad t \in [0, 1].$$

It can be shown that $M_n(\cdot)$ converges weakly to a Gauss-Markov process $M(\cdot)$ which satisfies the following equation

$$M(t) = \int_0^t e^{-(I-\Gamma H)(\ln u - \ln t)} d\hat{B}(u),$$

where $\hat{B}(\cdot)$ is a Brownian motion with covariance matrix $S(\Gamma)$. In addition,

$$\sqrt{n}(x_n - \theta) = M_n(1) + o(1)$$

where $o(1) \xrightarrow{n} 0$ in probability. Thus, $\sqrt{n}(x_n - \theta) \sim N(0, S)$, with $S(\Gamma)$ given by

$$S(\Gamma) = \int_0^\infty e^{(\frac{I}{2}-\Gamma H)u} \Gamma S_0 \Gamma' e^{(\frac{I}{2}-\Gamma H)' u} du,$$

where S_0 is the error covariance matrix and will be specified in the sequel. The problem then can be stated as that choose Γ so as to minimize $S(\Gamma)$. By means of algebraic method (cf. Wei [17, Theorem 1]), it can be shown that $S(\Gamma)$ is minimized when $\Gamma = \Gamma^* = -H^{-1} = -(f_x(\theta))^{-1}$ and the optimal covariance matrix is given by $S(\Gamma^*) = H^{-1} S_0 (H^{-1})'$. Note that this step does not depends on the particular structure of the noise process due to the fact it is purely algebraic. Because Γ is rarely available, one then constructs a sequence of estimates $\{\Gamma_n\}$ and uses the following algorithm instead

$$x_{n+1} = x_n + \frac{\Gamma_n}{n}(f(x_n) + \xi_n).$$

It can be established that as $n \to \infty$, $x_n \to \theta$ w.p.1, $\Gamma_n \to \Gamma^*$ w.p.1 and the asymptotic covariance is the optimal one given by $S(\Gamma^*)$. However, in view of the above equation, in addition to estimating θ, one needs to approximates $f_x(\theta)$ as well. Due to the estimation of the matrix-valued elements, such algorithms are computationally very intensive, especially for very large dimensional systems.

In lieu of the above adaptive procedures, Polyak suggested to use the following recursive scheme. Choose a sequence of slowly varying gain $a_n = \frac{1}{n^\gamma}$ with $\frac{1}{2} < \gamma < 1$. In place of Eq. (2.1), use a pair of equations to complete the calculations. To be more specific, define

$$\bar{x}_n = \frac{1}{n} \sum_{j=1}^{n} x_j.$$

The recursive algorithm is given by:

$$x_{n+1} = x_n + \frac{1}{n^\gamma}(f(x_n) + \xi_n) \tag{2.2.1}$$

$$\bar{x}_{n+1} = \bar{x}_n - \frac{1}{n+1}\bar{x}_n + \frac{1}{n+1}x_{n+1}. \tag{2.2.2}$$

Under the i.i.d. assumption on $\{\xi_n\}$, it was proved in [10] that

$$E(\bar{x}_n - \theta)(\bar{x}_n - \theta)' = \frac{1}{n}S^* + o\left(\frac{1}{n}\right), \tag{2.3}$$

where S^* is the "optimal asymptotic covariance matrix". Hence, the algorithm is an efficient one.

Note that in the procedure given by (2.2), first, no matrix-valued iterates are needed. Secondly, both $\{x_n\}$ and $\{\bar{x}_n\}$ can be obtained recursively. It thus achieves the same goal of asymptotic optimality as the adaptive procedures without carrying out extensive and additional computations. As far as the estimation problem is concerned, (2.2) seems to be more preferable.

To implement stochastic recursive algorithms, one would like to have the iterates get to the neighborhood of the true parameter fast enough. This can be done by choosing large step size a_n. However, larger step size will result in 'larger' asymptotic covariance of the scaled sequence. Thus, if one uses (2.2.1) alone, the aforementioned goal will be achieved at the expense of slower convergence rate. The use of the averaging approach allows one to get to the vicinity of the true parameter faster. In the mean time, it keeps the asymptotic covariance to be the optimal one. It produces a 'squeezing effect' which forces the iterates get to the vicinity of θ without paying the price of increasing the asymptotic covariance matrix or slowing down the convergence speed.

We shall obtain a similar result using the methods of weak convergence for correlated φ-mixing type of noise processes.

The following assumptions will be needed in the sequel:

(A1) There is a unique θ such that $f(\theta) = 0$ and

$$f(x) = H(x - \theta) + g(x),$$

where $|g(x)| = O(|x - \theta|^2)$ and H is a stable matrix in that all of its eigenvalues have negative real parts. The function $f(\cdot)$ satisfies the following growth condition:

$$|f(x)| \leq K(1 + |x|) \text{ for some } K > 0.$$

In addition, there is a twice continuously differentiable function $V(\cdot) : \mathbf{R}^r \to \mathbf{R}$ such that, $V(x) \geq 0$, $|V_{xx}(\cdot)|$ is bounded, $V(x) \xrightarrow{|x| \to \infty} \infty$, and for some $\mu > 0$ and all $x \neq \theta$, $V_x'(x)f(x) < -\mu V(x)$.

(A2) $\{\xi_n\}$ is a stationary φ-mixing sequence (cf. [12]) such that

$$E\xi_n = 0, \; E|\xi_n|^{2+\delta} < \infty \text{ for some } \delta > 0.$$

For $m > 0$, let the mixing measure be defined by

$$\varphi(m) = \sup_{A \in \mathcal{F}^{n+m}} |P(A|\mathcal{F}_n) - P(A)|_{\frac{2+\delta}{1+\delta}}$$

where $\mathcal{F}_n = \sigma\{\xi_k; k \leq n\}$, $\mathcal{F}^n = \sigma\{\xi_k; k \geq n\}$, and

$$|z|_{\frac{2+\delta}{1+\delta}} = E^{\frac{1+\delta}{2+\delta}}|z|^{\frac{2+\delta}{1+\delta}}.$$

Suppose $\sum_m \varphi^{\frac{\delta}{1+\delta}}(m) < \infty$.

Remark: (A1) consists of a stability condition and assumptions on $f(\cdot)$. We assumed that the function $f(\cdot)$ grows at most linearly, which is a standard assumption on the local behavior of $f(x)$ and its growth (cf. [19, Section 2]). In fact, much weaker conditions can be imposed. The mixing measure in (A2) is defined in terms of the p-norm (with $p = (2 + \delta)/(1 + \delta)$). When $p = \infty$, it is known as measure of uniform mixing, and when $p = 1$, it is termed measure of strong mixing. Previously, uniform mixing was frequently used in the study of rates of convergence issues (cf. [3]). The mixing assumption used here seems to be slightly weaker.

Under (A1) and (A2), the algorithm is strongly convergent, i.e.,

$$P\left(\lim_n \bar{x}_n = \theta\right) = 1. \tag{2.4}$$

In view of the ordinary differential equation (ODE) approach in [1, Chapter 2] (see also [14, Section 2] and the references therein), to establish this fact, we need only verify that

$$\sum_{j=1}^{\infty} \frac{1}{j^\gamma} \xi_j \text{ converges w.p.1.} \tag{2.5}$$

To verify that (2.5) holds, we shall apply the following assertion:

Claim. *Let $\{W_n\}$ be a sequence of vector-valued random variables. Suppose that there is a sequence of non-negative integers $\{\rho(n)\}$ and a constant $K > 0$ such that*

$$|EW_i W_{i+j}| \leq K\rho(j)|W_i|_2|W_{i+j}|_2, \quad \text{for all } i, j \geq 1 \tag{2.6}$$

$$\sum_{j=1}^{\infty} \rho(j) < \infty \tag{2.7}$$

$$\sum_{j=1}^{\infty} (\ln^2 j) E|W_j|^2 < \infty. \tag{2.8}$$

Then, $\sum_{j=1}^{\infty} W_j$ converges w.p.1.

The proof of this claim can be obtained by using a modification of [20, Corollary 2.4.1]. Using the above claim, define $W_n = \frac{1}{j^\gamma}\xi_n$. Then, it is easily seen that (2.6) and (2.7) are satisfied with $\rho(n) = \varphi^{\frac{\delta}{1+\delta}}(n)$. In addition, it can be seen that

$$\sum_{j=1}^{\infty} (\ln^2 j) E|W_j|^2 = |\xi_1|_2^2 \sum_{j=1}^{\infty} \frac{(\ln^2 j)}{j^{2\gamma}} < \infty.$$

Thus, (2.8) also holds. As a result, (2.5) follows. By the use of (2.5) and a standard result of the ODE approach (cf. [1]), $x_n \xrightarrow{n} \theta$ w.p.1. Noticing the fact that $\bar{x}_n$ is the average of x_j, $j \leq n$, (2.4) then follows.

Remark: An alternative proof of the convergence can be obtained by defining:

$$M_n = \sum_{j=1}^{n} \frac{1}{j^\gamma} \xi_j + \sum_{m=1}^{\infty} \frac{1}{(n+m)^\gamma} E\left(\xi_{n+m} | \mathcal{F}_n\right),$$

and $m_n = M_n - M_{n-1}$. It can be shown that the second term on the right-hand side of the definition of M_n converges w.p.1. We then prove the convergence of M_n by using detailed estimates via mixing inequalities and martingale convergence theorem. As a consequence, (2.5) can be established.

For simplicity and without loss of generality, we shall assume $\theta = 0$ whenever is convenient. To proceed, we state a preparatory lemma first.

Lemma 2.1. *Suppose the conditions (A1)-(A2) are satisfied. Then, for sufficiently large n, $EV(x_n) = O(n^{-\gamma})$.*

Proof: We prove the assertion by using the techniques of perturbed Liapunov functions (cf. [3]). Let E_n denote the conditional expectation on the σ-algebra $\{\xi_j; j \leq n\}$. We have that

$$E_{n-1} V(x_{n+1}) - V(x_n) = \frac{1}{n^\gamma} V_x'(x) f(x_n) + \frac{1}{n^\gamma} V_x'(x_n) E_{n-1} \xi_n + O\left(\frac{1}{n^{2\gamma}}\right) + \zeta_n, \qquad (2.9)$$

where $E\zeta_n = O\left(n^{-2\gamma}\right)$. Define

$$W(x, n) = \sum_{j=n}^{\infty} \frac{1}{j^\gamma} V_x'(x) E_{n-1} \xi_j.$$

In view of Corollary 7.2.4 in [12],

$$\sum_{j=n}^{\infty} E|E_{n-1}\xi_j| \leq 8 \sum_{j=n}^{\infty} \varphi^{\frac{4}{1+\delta}}(j-n)|\xi_1|_{2+\delta} < \infty.$$

Thus,

$$E|W(x, n)| \leq \frac{K}{n^\gamma} |V_x'(x)| \leq \frac{K}{n^\gamma}(1 + V(x)). \qquad (2.10)$$

Define

$$\tilde{V}(x, n) = V(x) + W(x, n).$$

Direct computation yields

$$E_{n-1} \tilde{V}(x_{n+1}, n+1) - \tilde{V}(x_n, n) \leq -\frac{\mu}{n^\gamma} \tilde{V}(x_n, n) + O\left(\frac{1}{n^{2\gamma}}\right) + \tilde{\zeta}_n,$$

where $E\tilde{\zeta}_n = O\left(n^{-2\gamma}\right)$. Taking expectations, and iterating on the resulting inequality, we obtain that for some M,

$$E\tilde{V}(x_{n+1}, n+1) \leq \prod_{j=M}^{n} \left(1 - \frac{\mu}{j^\gamma}\right) E\tilde{V}(x_M, M) + K \sum_{k=M}^{n} \prod_{j=k+1}^{n} \left(1 - \frac{\mu}{j^\gamma}\right) \frac{1}{k^{2\gamma}}. \qquad (2.11)$$

It is easily seen that

$$\prod_{j=k+1}^{n} \left(1 - \frac{\mu}{j^\gamma}\right) \leq \exp\left(-\mu \sum_{j=k+1}^{n} \frac{1}{j^\gamma}\right).$$

In view of the above inequality, to get the desired order of magnitude estimate, we need only examine the second term on the right-hand side of (2.11). Notice that

$$\sum_{k=1}^{n} \prod_{j=k+1}^{n} \left(1 - \frac{\mu}{j^\gamma}\right) \frac{1}{k^{2\gamma}} = \frac{1}{n^\gamma} \sum_{k=1}^{n} \frac{1}{k^\gamma} \prod_{j=k+1}^{n} \left(1 - \frac{\mu}{j^\gamma}\right)$$

$$+ \sum_{k=1}^{n-1} \left(\frac{1}{k^\gamma} - \frac{1}{(k+1)^\gamma}\right) \left(\sum_{l=1}^{k} \frac{1}{l^\gamma} \prod_{j=l+1}^{n} \left(1 - \frac{\mu}{j^\gamma}\right)\right). \quad (2.12)$$

We obtain that

$$\sum_{k=1}^{n} \frac{1}{k^\gamma} \prod_{j=k+1}^{n} \left(1 - \frac{\mu}{j^\gamma}\right) = \frac{1}{\mu} \sum_{k=1}^{n} \left(\prod_{j=k+1}^{n} \left(1 - \frac{\mu}{j^\gamma}\right) - \prod_{j=k}^{n} \left(1 - \frac{\mu}{j^\gamma}\right)\right)$$

$$= \frac{1}{\mu} \left(1 - \prod_{j=1}^{n} \left(1 - \frac{\mu}{j^\gamma}\right)\right) \leq \frac{1}{\mu}. \quad (2.13)$$

Using (2.13), it can be shown that the first term on the right-hand side of (2.12) is of the order $O(n^{-\gamma})$. As for the second term, by virtue of a Taylor expansion,

$$\frac{1}{(k+1)^\gamma} = \frac{1}{k^\gamma} \left(1 - \gamma \frac{1}{k} + O\left(\frac{1}{k^2}\right)\right).$$

Noticing the fact

$$\sum_{k=1}^{n} \frac{1}{k^{1+\gamma}} = O(n^{-\gamma}),$$

we have

$$E\tilde{V}(x_n, n) = O(n^{-\gamma}) \text{ for sufficiently large } n.$$

This and (2.10) then yield the desired result. □

Remark: If $V(x) = x'Ax + o(|x|^2)$, where A is a positive definite matrix, then the above lemma implies that for n large enough,

$$E|x_n|^2 = O(n^{-\gamma}). \quad (2.14)$$

Note that (2.14) is only needed in the following asymptotic normality result. It is not needed for the almost sure convergence of the algorithm. We are now ready to state the main theorem.

Theorem 2.2. Let $S = H^{-1}S_0(H^{-1})'$ with

$$S_0 = E(\xi_1\xi_1') + \sum_{k=2}^{\infty} E(\xi_1\xi_k') + \sum_{k=2}^{\infty} E(\xi_k\xi_1'). \quad (2.15)$$

Under the assumptions (A1), (A2) and (2.14),

$$\sqrt{n}(\bar{x}_n - \theta) \xrightarrow{n} N(0, S) \text{ in distribution.}$$

Remark: The infinite series in (2.15) are absolutely convergent due to the mixing inequalities used previously. In view of the discussion in the introduction section, Theorem 2.2 above indicates that the algorithm (2.2) has the optimal convergence rate in that the asymptotic covariance matrix is the optimal one.

3. Proof of Theorem 2.2

Define

$$A_{nj} = \begin{cases} \prod_{k=j+1}^{n}(I + H/k^\gamma), & n \geq j+1; \\ I; & n = j. \end{cases}$$

In view of (2.2), for any $m > 0$,

$$x_{n+1} = A_{n,m-1}x_m + \sum_{j=m}^{n} \frac{1}{j^\gamma} A_{nj} g(x_j) + \sum_{j=m}^{n} \frac{1}{j^\gamma} A_{nj} \xi_j, \tag{3.1}$$

and

$$\sqrt{n+1}\bar{x}_{n+1} = \frac{1}{\sqrt{n+1}} \sum_{k=1}^{m-1} x_k + \frac{1}{\sqrt{n+1}} \sum_{k=m}^{n} A_{k,m-1}x_m$$

$$+ \frac{1}{\sqrt{n+1}} \sum_{k=m}^{n} \sum_{j=m}^{k} \frac{1}{j^\gamma} A_{kj} g(x_j) + \frac{1}{\sqrt{n+1}} \sum_{k=m}^{n} \sum_{j=m}^{k} \frac{1}{j^\gamma} A_{kj} \xi_j. \tag{3.2}$$

Notice that

$$|A_{nj}| \leq \exp\left(-\lambda \sum_{k=j+1}^{n} k^{-\gamma}\right)$$

for some $\lambda > 0$. In what follows, we choose a sequence so that $m(n) \xrightarrow{n} \infty$ but $m(n)/\sqrt{n} \xrightarrow{n} 0$. For simplicity, we shall write m for $m(n)$ in the sequel. The next lemma indicates that in order to study the asymptotic normality, we need only consider the last term on the right-hand side of (3.2).

Lemma 3.1. *Under the conditions of Theorem 2.2,*

$$\frac{1}{\sqrt{n+1}} \sum_{k=1}^{m-1} x_k \xrightarrow{n} 0 \ w.p.1$$

$$\frac{1}{\sqrt{n+1}} \sum_{k=m}^{n} A_{k,m-1}x_m \xrightarrow{n} 0 \ in \ probability$$

$$\frac{1}{\sqrt{n+1}} \sum_{k=m}^{n} \sum_{j=m}^{k} \frac{1}{j^\gamma} A_{kj} g(x_j) \xrightarrow{n} 0 \ in \ probability.$$

Proof: The first equation above can be verified by noticing the almost sure boundedness of $\{x_k\}$, whereas the second equation can be proved by observing that

$$E\left| \frac{1}{\sqrt{n+1}} \sum_{k=m}^{n} A_{k,m-1}x_m \right|$$

$$\leq K \frac{1}{\sqrt{n+1}} \sum_{k=m}^{n} \exp\left(-\frac{\lambda}{1-\gamma}(k^{1-\gamma} - m^{1-\gamma})\right) \xrightarrow{n} 0.$$

As for the last equation, owing to the Markov inequality, for any $\eta > 0$,

$$P\left\{\left|\frac{1}{\sqrt{n+1}}\sum_{k=m}^{n}\sum_{j=m}^{k}\frac{1}{j^\gamma}A_{kj}g(x_j)\right| > \eta\right\}$$

$$\leq \frac{K}{\eta}\frac{1}{\sqrt{n+1}}\sum_{k=m}^{n}\sum_{j=m}^{k}\frac{1}{j^\gamma}|A_{kj}|E|x_j|^2 \xrightarrow{n} 0.$$

The lemma is thus proved. $\square$

Lemma 3.2 below shows that the study of the asymptotic distribution of the third term on the right-hand side of (3.2) can further be reduced to $-\frac{1}{\sqrt{n}}H^{-1}\sum_{j=1}^{n}\xi_j$.

Lemma 3.2. *Under the conditions of Theorem 2.2,*

$$T_n = \frac{1}{\sqrt{n+1}}\sum_{k=m}^{n}\sum_{j=m}^{k}\frac{1}{j^\gamma}A_{kj}\xi_j = -\frac{H^{-1}}{\sqrt{n}}\sum_{j=1}^{n}\xi_j + o(1), \tag{3.3}$$

where $o(1)\xrightarrow{n}0$ *in probability.*

Proof: Interchanging the order of summations and noticing the fact $\frac{1}{\sqrt{n+1}} - \frac{1}{\sqrt{n}}\xrightarrow{n}0$ yield that

$$T_n = \frac{1}{\sqrt{n}}\sum_{j=m}^{n}\frac{1}{j^\gamma}\left(\sum_{k=j}^{n}A_{kj}\right)\xi_j + o(1). \tag{3.4}$$

We observe that for $j \geq m$,

$$\ln A_{kj} = \sum_{l=j+1}^{k}\ln(I + H/l^\gamma)$$

$$= \sum_{l=j+1}^{k}\left(H/l^\gamma - H^2/(2l^{2\gamma}) + \cdots\right). \tag{3.5}$$

Hence,

$$A_{kj} = \exp\left(H\sum_{l=j+1}^{k}l^{-\gamma}\right)O\left(\exp\left(H^2\sum_{l=j+1}^{k}l^{-2\gamma}\right)\right)$$

$$= \exp\left(H\sum_{l=j+1}^{k}l^{-\gamma}\right)I_{kj}, \tag{3.6}$$

where $I_{kj} \to I$, the identity matrix as $j \to \infty$ uniformly in $k > j$. In view of (3.4) and (3.6), the 'effective' terms in T_n is given by:

$$\tilde{T}_n = \frac{1}{\sqrt{n}}\sum_{j=m}^{n}\frac{1}{j^\gamma}\sum_{k=j}^{n}\exp\left(H\sum_{l=j+1}^{k}l^{-\gamma}\right),$$

or equivalently, $T_n = \tilde{T}_n + o(1)$, where $o(1)\xrightarrow{n}0$ in probability.

Without loss of generality, we may assume $n \geq j + j^{2\gamma}$. Then, the limit (in distribution) of $\tilde{T}_n$ is the same as that of

$$\frac{1}{\sqrt{n}} \sum_{j=m}^{n} \frac{1}{j^\gamma} \left\{ \int_{j}^{j+j^{2\gamma}} \exp\left(Hj^{-\gamma}(x-j)\right) dx \right\} \xi_j$$

$$+ \frac{1}{\sqrt{n}} \sum_{j=m}^{n} \frac{1}{j^\gamma} \sum_{k=j+j^{2\gamma}}^{n} \exp\left(H \sum_{l=j+1}^{k} l^{-\gamma} \right) \xi_j. \tag{3.7}$$

We claim that:

(a) the first term in (3.7) has the same limit in distribution as that of the term $-H^{-1}/\sqrt{n} \sum_{j=m}^{n} \xi_j$.

(b) the second term in (3.7) tends to 0 in probability.

A straight forward integration yields that

$$\frac{1}{j^\gamma} \int_{j}^{j+j^{2\gamma}} \exp(Hj^{-\gamma}(x-j)) dx = H^{-1}\left(\exp(Hj^\gamma) - I\right).$$

In addition, for some $\lambda > 0$,

$$E \left| \frac{1}{\sqrt{n}} \sum_{j=m}^{n} \exp(Hj^\gamma)\xi_j \right|$$

$$\leq K \frac{1}{\sqrt{n}} \sum_{j=m}^{n} \exp(-\lambda j^\gamma) \leq K \frac{1}{\sqrt{n}} \sum_{j=m}^{n} \frac{1}{1+\lambda j^\gamma}$$

$$= O\left(n^{\frac{1}{2}-\gamma}\right) \xrightarrow{n} 0. \tag{3.8}$$

Thus, assertion (a) is verified. Using some detailed estimates, (b) can also be proved.

Finally, we note that $\frac{1}{\sqrt{n}} \sum_{j=1}^{m} \xi \xrightarrow{n} 0$ in probability by the choice of $m(n)$, and hence

$$-\frac{H^{-1}}{\sqrt{n}} \sum_{j=1}^{n} \xi_j = -\frac{H^{-1}}{\sqrt{n}} \sum_{j=m}^{n} \xi_j + o(1),$$

where $o(1) \xrightarrow{n} 0$ in probability. Lemma 3.3 thus follows. $\square$

Define

$$B_n(t) = \frac{1}{\sqrt{n}} \sum_{j=1}^{[nt]} \xi_j$$

$$\tilde{B}_n(t) = \frac{1}{\sqrt{n}} \sum_{j=1}^{[nt]} x_j,$$

where $[z]$ denotes the largest integral part of z.

Lemma 3.3. *Under the conditions of Theorem 2.2,*

(1) $B_n(\cdot)$ *converges weakly to a Brownian motion $B(\cdot)$ with covariance matrix S_0, where S_0 is given by (2.15).*

(2) $\tilde{B}_n(\cdot)$ *converges weakly to a Brownian motion* $\tilde{B}(\cdot)$ *with covariance matrix*

$$S = H^{-1}S_0(H^{-1})'.$$

Proof: (1) is a multidimensional extension of a corresponding result in [12]. In view of Lemma 3.1 and Lemma 3.2, $\tilde{B}_n(t) = -H^{-1}B_n(t) + o(1)$ where $o(1)\xrightarrow{n}0$ in probability. As a result, (2) follows from (1) by virtue of the well-known Slutsky's Theorem. $\square$

Remark: It follows from the above lemma, the covariance functions for $B(\cdot)$ and $\tilde{B}(\cdot)$ are $\Sigma(t) = S_0t$ and $\tilde{\Sigma}(t) = St$, respectively.

Proof of Theorem 2.2: Setting $t = 1$ in $\tilde{B}(t)$ and using Lemma 3.1 and 3.2, the desired asymptotic normality follows. Hence the proof is completed. $\square$

4. Examples

We begin this section by examining two simple examples. Our objective is to compare the asymptotic performance of different algorithms. For simplicity, we shall consider scalar problems only. Several different random processes are dealt with. They include, i.i.d. random sequences, moving average process driven by a white noise sequence and autoregressive moving average models with order (1,1) (i.e., ARMA(1,1)).

A linear function

$$f(x) = 2(6 - x), \tag{4.1}$$

and a nonlinear function

$$f(x) = 4.2 - 2x - 0.005(x - 2.1)^3 \tag{4.2}$$

will be treated in the two examples, respectively.

We shall carry out the computation by using a standard stochastic approximation algorithm, a stochastic approximation algorithm with optimal gain, and a two-stage averaging procedure. To make the comparisons, we compute the sample variances for the classical approach and that of the averaging algorithm.

Table 4.1-4.6 list some of the numerical results concerning their sample variances. Table 4.1-4.3 are the results corresponding to (4.1), whereas Table 4.4-4.6 deal with (4.2).

It is well-known that for the single-stage or classical approach, the best choice of γ (with $a_n = 1/n^\gamma$, $0 < \gamma \le 1$) is $\gamma = 1$. As a result, we compare the performance of the averaging procedure vs. that of the classical algorithm with $a_n = 1/n$. For the purposes of illustration, we also make comparisons between the averaging algorithm and the algorithm having 'optimal' gain, i.e., the constant a in $a_n = a/n$ is the optimal choice. Since we are mainly interested in the performance of these algorithms, only the sample variances are listed in the tables.

The choice of a_n in the calculation of $\tilde{x}_n$ is the bast possible one, i.e., a is optimal which leads to the optimality of the performance of the algorithm. Such an a is normally not known. For comparison reasons, we put the results here, however.

Table 4.1 Algorithms with $x_1 = 4.5$, $\gamma = 0.8$, i.i.d. noise

Iter. no.	MSE for $\hat{x}_n$	MSE for $\tilde{x}_n$	MSE for $\bar{x}_n$
300	.0405703300	.0118087500	.0113226400
500	.0244228000	.0071168860	.0068192880
800	.0154764000	.0045213180	.0043220720

Table 4.2 Algorithms with $x_1 = 4.5$, $\gamma = 0.8$, MA noise

Iter. no.	MSE for $\hat{x}_n$	MSE for $\tilde{x}_n$	MSE for $\bar{x}_n$
300	.0325413100	.0140296700	.0124101300
500	.0198818900	.0085536670	.0074621230
800	.0136115700	.0056569100	.0047978950

Table 4.3 Algorithms with $x_1 = 4.5$, $\gamma = 0.8$, ARMA noise

Iter. no.	MSE for $\hat{x}_n$	MSE for $\tilde{x}_n$	MSE for $\bar{x}_n$
300	.0280318800	.0125388500	.0113027500
500	.0170574400	.0076202920	.0067873290
800	.0116089700	.0049599090	.0043828290

Table 4.4 Algorithms with $x_1 = 0$, $\gamma = 0.55$, i.i.d. noise

Iter. no.	MSE for $\hat{x}_n$	MSE for $\tilde{x}_n$	MSE for $\bar{x}_n$
300	.0619785400	.0194125200	.0181010400
500	.0372424200	.0116645000	.0108748900
800	.0235159900	.0073631700	.0068494500

In these tables, the first column lists the iteration number; the second column gives the computed sample variance for $\hat{x}_n$ of the classical algorithm with $a_n = 1/n$; the third column gives the computed sample variance of $\tilde{x}_n$ of the algorithm with $a_n = a/n$, where $a = -H^{-1}$; the last column gives sample variance for $\bar{x}_n$.

Table 4.5 Algorithms with $x_1 = 0$, $\gamma = 0.55$, MA noise

Iter. no.	MSE for $\hat{x}_n$	MSE for $\tilde{x}_n$	MSE for $\bar{x}_n$
300	.0396919500	.0188987100	.0193427900
500	.0240001000	.0114396800	.0117604200
800	.0156303200	.0073074010	.0075143760

Table 4.6 Algorithms with $x_1 = 0$, $\gamma = 0.55$, ARMA noise

Iter. no.	MSE for $\hat{x}_n$	MSE for $\tilde{x}_n$	MSE for $\bar{x}_n$
300	.0393224800	.0187159000	.0184650200
500	.0237470800	.0112951000	.0111834900
800	.0154295100	.0072092330	.0071326700

In Table 4.1, the noise is an i.i.d. sequence; the random disturbance in Table 4.2 is of the form

$$\xi_i = w_i + 0.6w_{i-1} + 0.7w_{i-2}; \tag{4.3}$$

finally, an ARMA process

$$\xi_i = 0.4\xi_{i-1} + w_i + 0.2w_{i-1} \tag{4.4}$$

is treated in Table 4.3.

Similarly, for the nonlinear functions, i.i.d. noise, a MA process of the form

$$\xi_i = w_i + 0.4w_{i-1} + 0.3w_{i-2} \tag{4.5}$$

and an ARMA model

$$\xi_i = 0.2\xi_{i-1} + w_i + 0.3w_{i-1} \tag{4.6}$$

are considered. The corresponding numerical results are tabulated in Table 4.4 through 4.6.

Remark: In the second example, the condition on linear growth is actually violated. To overcome the possible problems, we have used a simple truncation device (or restarting device) in the computation. In practice, one will probably use a projection or truncation algorithm to carry out the computation any way. There many of them are available.

The MA sequences considered in the examples are m-dependent processes which satisfy the mixing conditions. The ARMA models considered here are functions of mixing processes (cf. [21]). Using a slight modification of the arguments in this paper, it can be shown that the main results presented here still hold.

It can easily be seen from the above tables, the improvement of the performance of the averaging algorithm is demonstrated when the iteration number is large. If n is large enough, then the variance of the averaged approach is much smaller.

In view of the comparisons between the algorithms with optimal gain and the averaging approach, the sample variances of $\bar{x}_n$ are better than that of $\tilde{x}_n$ in most of the time. Of course, care must be taken to interpret these results since it is not clear what 'asymptotic' means if the number of iterations is only a few hundreds or even less.

In this section, only simple scalar problems are dealt with. The advantages of the averaging algorithms will particularly be pronounced for multidimensional systems. It provides significant savings for the memory storage and the amount of computations.

5. Further discussions

By proving a weak convergence theorem, we have shown that the multistage algorithm via averaging is an efficient procedure in that it has the optimal rate of convergence.

The algorithm (2.2.1)-(2.2.2) calls an averaging procedure starting from the first iteration. In the actual implementation, one may wish to modify this approach. For instance, one could begin the averaging after a number of iterations so that the procedure has passed 'transient' period and reached the neighborhood of the true parameter. Thus, one may define $\{x_n\}$ as (2.2.1) and let

$$\bar{x}_n = \frac{1}{n} \sum_{j=\mu}^{\mu+n} x_j, \text{ for some } \mu. \tag{5.1}$$

Due to the stationarity of $\{\xi_n\}$, the asymptotic optimality result derived previously still holds. We get $\sqrt{n}\bar{x}_n \sim N(0, S)$ with $S = H^{-1}S_0(H^{-1})'$ and S_0 given in (2.15) (Again, we assumed that $\theta = 0$ for convenience).

In addition, 'moving windows' can also be constructed. In lieu of (2.2.2) or (5.1), we can define

$$\bar{x}_n^{p_n,q_n} = \frac{1}{q_n} \sum_{j=p_n}^{p_n+q_n} x_j, \tag{5.2}$$

and

$$\bar{B}_n^{p_n,q_n}(t) = \frac{1}{\sqrt{q_n}} \sum_{j=p_n}^{p_n+[q_n t]} x_j, \tag{5.3}$$

where $[z]$ denotes the largest integral part of z and $\{p_n\}$ $\{q_n\}$ are sequences of positive integers satisfying $q_n \xrightarrow{n} \infty$. Using conditions of Theorem 2.2, the same argument as before yields that

$$\bar{B}_n^{p_n,q_n}(t) = -\frac{H^{-1}}{\sqrt{q_n}} \sum_{j=1}^{[q_n t]} \xi_j + o(1) \Rightarrow \tilde{B}(t), \tag{5.4}$$

where $\tilde{B}(\cdot)$ is as in Lemma 3.3. Hence, again the asymptotic optimality holds. The above idea of 'moving window' was suggested in Kushner and Yang [22] with $p_n = n$. (5.4) above indicates that neither the starting point of the averaging nor the size of the window is crucial.

In this paper, only algorithms with additive noise is discussed. Similar idea can be employed to treat algorithms of the form

$$x_{n+1} = x_n + a_n f(x_n, \xi_n). \tag{5.5}$$

We can design an procedure via averaging by constructing:

$$x_{n+1} = x_n + \frac{1}{n^\gamma} f(x_n, \xi_n), \text{ for } \frac{1}{2} < \gamma < 1$$

$$\bar{x}_{n+1} = \bar{x}_n - \frac{1}{n+1}\bar{x}_n + \frac{1}{n+1}x_{n+1}.$$

(5.6)

The proof follows along similar line as the argument presented in the preceding sections. However, modifications and care must be taken due to the non-additive noise processes.

Furthermore, the averaging approach may be employed in various recursive stochastic algorithms (such as gradient algorithm, quasigradient methods, their variations and various projection procedures) for optimization, identification and adaptive signal processing. It is conceivable that the resulting algorithms will enable one to obtain robust estimates and solve the underlying problems more efficiently.

References

[1] H.J. Kushner and D.S. Clark, *Stochastic Approximation for Constrained and Unconstrained Systems*, Springer-Verlag, Berlin, 1978.

[2] G. Ch. Pflug, Stochastic minimization with constant step-size: asymptotic laws, *SIAM J. Control Optim.*, **24** (1986), 655-666.

[3] H.J. Kushner, *Approximation and Weak Convergence Methods for Random Processes with Applications to Stochastic Systems Theory*, MIT Press, Cambridge, MA, 1984.

[4] J.N. Tsitsiklis, D.P. Bertsekas and M. Athans, Distributed asynchronous deterministic and stochastic gradient optimization algorithms, *IEEE Trans. Automat. Control*, AC-31 (1986), 803-812.

[5] D.P. Bertsekas and J.N. Tsitsiklis, *Parallel and Distributed Computing*, Prentice-Hall, New Jersey, 1989.

[6] H.J. Kushner and G. Yin, Asymptotic properties of distributed and communicating stochastic approximation algorithms, *SIAM J. Control Optim.*, **25** (1987), 1266-1290.

[7] H.J. Kushner and G. Yin, Stochastic approximation algorithms for parallel and distributed processing, *Stochastics*, **22** (1987), 219-250.

[8] G. Yin and Y.M. Zhu, On w.p.1 convergence of a parallel stochastic approximation algorithm, *Probab. Eng. Inform. Sci.*, **3** (1989), 55-75.

[9] G. Yin, Recent progress in parallel stochastic approximations, to appear in *Statistical Theory of Identification and Adaptive Control*, (P.E. Caines and L. Gerencsér Eds.), Springer-Verlag, 1991.

[10] B.T. Polyak, New method of stochastic approximation type, *Automat. Remote Control* **51** (1990), 937-946.

[11] D. Ruppert, Efficient estimations from a slowly convergent Robbins-Monro process, Technical Report, No. 781, School of Oper. Res. & Industrial Eng., Cornell Univ., 1988.

[12] S.N. Ethier and T.G. Kurtz, *Markov Processes, Characterization and Convergence*, Wiley, New York, 1986.

[13] A. Benveniste, M. Metivier and P. Priouret, *Adaptive Algorithms and Stochastic Approximation*, Springer-Verlag, Berlin, 1990.

[14] K.L. Chung, On a stochastic approximation method, *Ann. Math. Statist.* **25** (1954), 463-483.

[15] J.H. Venter, An extension of the Robbins-Monro procedure, *Ann. Math. Statist.* **38** (1967), 181-190.

[16] T.L. Lai and H. Robbins, Consistency and asymptotic efficiency of slope estimates in stochastic approximation schemes, *Z. Wahrsch. verw. Gebiete* **56** (1981), 329-360.

[17] C.Z. Wei, Multivariate adaptive stochastic approximation, *Ann. Statist.* **15** (1987), 1115-1130.

[18] G. Yin, A stopping rule for the Robbins-Monro method, *J. Optim. Theory Appl.*, **67** (1990), 151-173.

[19] H.J. Kushner, Stochastic approximation with discontinuous dynamics and state dependent noise: w.p.1 and weak convergence, *J. Math. Anal. Appl.* **82** (1981), 527-542.

[20] W.F. Stout, *Almost sure convergence*, Academic Press, London, 1974.

[21] P. Billingsley, *Convergence of Probability Measure*, wiley, New York, 1968.

[22] H.J. Kushner and J. Yang, manuscript in preparation, 1991.

Nonuniform Random Numbers: A Sensitivity Analysis for Transformation Methods

Lothar Afflerbach

Institut für Statistik, TU Graz

Lessingstr. 27, A–8010 Graz

Wolfgang Hörmann

Institut für Statistik, WU Wien

Augasse 2–6, A–1090 Wien

Abstract. There are many methods for the transformation of uniform random numbers into nonuniform random numbers. These methods are employed for pseudo-random numbers generated by computer programs. It is shown that the sensitivity to the pseudo-random numbers used can vary a lot between the transformation methods. A classification of the sensitivity of several transformation methods is given. Numerical examples are presented for various transformation methods.

1 Introduction

Nowadays mostly pseudo-random numbers are used instead of truely random numbers. The most popular method of generating pseudo-random numbers is the linear congruential method. A **linear congruential generator (LCG)** [31] produces a sequence $\{x_i\}$ of m different integers with an integral initial value x_0 by the recurrence

$$x_i \equiv a \cdot x_{i-1} + b \pmod{m}, \; 0 \leq x_i \leq m-1, \; i = 1,2,3,\ldots, \tag{1}$$

if the three integers m (modulus), a (multiplier) and b (increment) are properly chosen as will be assumed in the following (see [30], Chapter 3.2, Theorem A). The fractions $u_i = x_i/m$ are used as random numbers uniformly distributed on $[0,1)$.

If n successively generated numbers $u_i, u_{i+1}, \ldots, u_{i+n-1}$ are to be taken as realisations of n independent random variables uniformly distributed on $[0,1)$ the equidistribution on the set of all n-dimensional vectors (formed by n successively generated numbers) should approximate the n-dimensional Lebesgue measure on $[0,1)^n$ for $n = 2,3,4,\ldots$ as well as possible. As is well known the set of the generated points has a lattice structure (see [13], [19], [21], [33], [34] and [1], [2], [40] for the sub-lattice structure of the non-overlapping vectors).

The lattice structure of linear congruential generators can be assessed via reduced lattice bases, hyperplane structure, the spectral test or the discrepancy (see [3] for a comparison of these criteria). Here we consider the ratio of the lengths of the shortest and the largest vector of a Minkowski reduced lattice basis for the assessment of the n-dimensional lattice structure [5]. This ratio, called Beyer quotient [13], has a value near 1 for a good lattice structure (all basis vectors have nearly the same lengths). For a bad lattice stucture the ratio is nearby 0.

In simulations several non-uniform distributions have to be realized. Many methods for generating non-uniform random numbers are presented by Ahrens and Dieter [10] and Devroye [18]. Investigations about the quality of transformed random numbers with respect to the used uniform pseudo-random numbers are hard to find.

In 1973 Neave examined the Box-Muller method (for the normal distribution) in connection with a multiplicative congruential generator with a very bad two-dimensional distribution of the generated pair [37]. For a special sample Neave found some deficiencies in the frequency distribution of one of the coordinates. Thus Neave warned not to use the Box-Muller method with linear congruential generators. Two years later Chay, Fardo und Mazumdar [17] proposed to exchange U_i and U_{i+1} in oder to have no deficiencies in the special example treated by Neave. In 1983 the discussion of Bratley, Fox and Schrage [16] of the Box-Muller method with linear congruential generators yields the same warning of this combination. In addition they pointed out that the two-dimensional numbers (x_i, x_{i+1}) lie on a spiral and they presented a figure with a small spiral. This spiral would belong to a generator for which the pairs are lying on only 5 lines in the unit square. This follows from the analysis of the spiral structure made by Afflerbach and Wenzel in [7]. In that paper the conjunction of linear congruential generators with the Box-Muller method and with the polar method was studied in more detail. For several transformation methods some nice dotplots of two-dimensional normal distributed pseudo-random numbers are shown by Ripley who thought *"There is little to choose between the Box-Muller, polar and ratio methods for sensitivity to pseudo-random number generators."* (p. 82 in [40]). In 1989 Afflerbach showed in [4] that the sensitivity to the pseudo-random numbers used can vary a lot between the transformation methods. This leads to a classification of the transformation methods.

2 Sensitivity analysis of transformation methods

In this section we examine several transformation methods with respect to special deficiencies in the distribution of the uniform random numbers in $2, 3, 4, \ldots$ dimensions. We consider two (small) linear congruential generators (1): The LCG 1 with $m = 16384$, $a = 129$ and $b = 1$ has a very good distibution of pairs – Beyer quotient $q_2 \approx 0.98$ (0.9999) for the lattice (sub-lattice) – but bad values $q_n \approx 0.03$ for $n = 3, 4, \ldots, 10$. The LCG 2 with $m = 16384$, $a = 9101$ and $b = 1$ has bad values in two dimensions but good values in $4, 5, 6, \ldots$ dimensions: $q_2 \approx 0.01$ (0.02), $q_3 \approx 0.13$, $q_4 \approx 0.77$ (0.47), $q_5 \approx 0.77$, $q_6 \approx 0.59$ (0.68), $q_7 \approx 0.73$, $\ldots$. Now we show that some of the transformation methods are more and some are less susceptible to the deficiencies in pesudo-random number generators.

In applications the normal distribution is of prime importance. A very often used method for the generation of normally distributed (pseudo-) random numbers is the following method [15].

Method 1: *The* **Box-Muller method** *transforms two independent, (0,1)-uniformly distributed random variables U_1, U_2 into the two independent, normally distributed random variables Y_1 and Y_2 (with mean 0 and standard deviation 1) by*

$$Y_1 = \sqrt{-2 \cdot \ln U_1} \cdot \cos(2 \cdot \pi \cdot U_2) \qquad \text{and} \qquad Y_2 = \sqrt{-2 \cdot \ln U_1} \cdot \sin(2 \cdot \pi \cdot U_2).$$

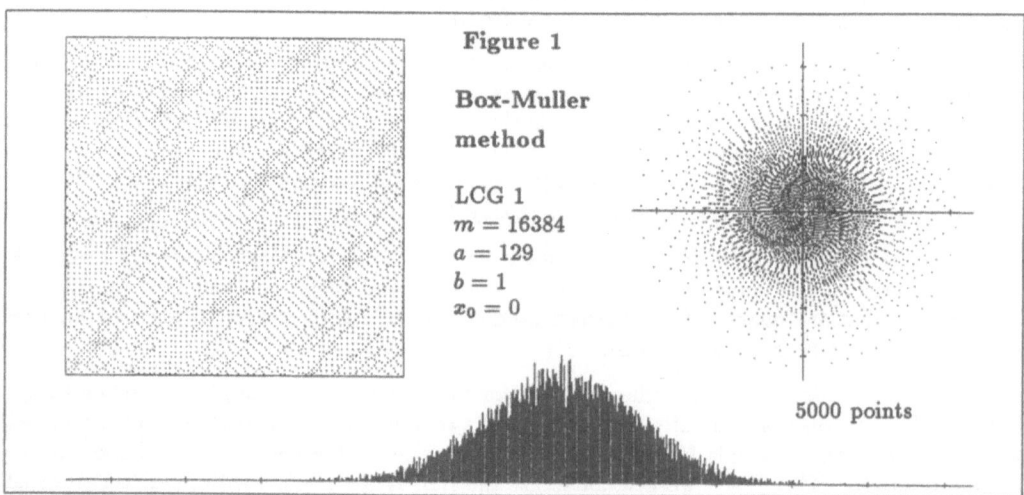

Figure 1

Box-Muller
method

LCG 1
$m = 16384$
$a = 129$
$b = 1$
$x_0 = 0$

5000 points

Remark 1: *The Box-Muller method is a two-to-two method. We have a good distribution of the transform numbers in two dimensions if and only if we have a good two-dimensional distribution of the uniform random numbers used. The lattice structure of a linear congrential generator is transformed into a spiral structure. This was studied in detail in [7]. Figure 1 shows as an example 5000 generated pairs of LCG 1 (good in 2, bad in 3, 4, . . . dimensions) and the transformed normally distributed pairs as well as a histogram of the one-dimensinal distribution of the normal numbers. The generated pairs occur in strange wave patterns because of the bad distribution in 4, 6, . . . dimensions. As it is shown in figure 2 deficiencies in uniform pseudo-random numbers in two dimensions generated by LCG 2 (which is bad in 2, acceptable in 3 and good in 4, 5, . . . dimensions) can even yield deficiencies in the one-dimensional distribution of the transformed numbers. But for all even numbers k the distribution of the uniform pseudo-random numbers in k dimensions corresponds in a one-to-one way to the k-dimensinal distribution of the transformed numbers.*

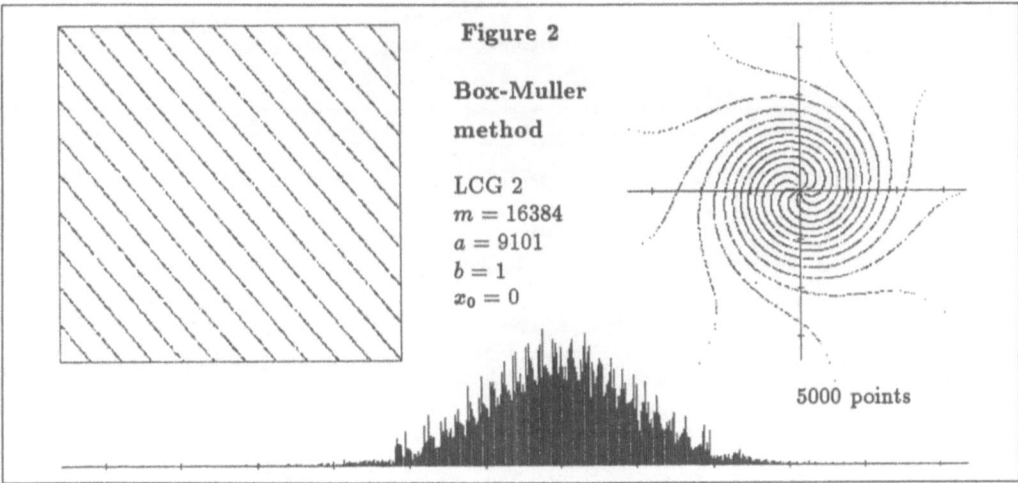

Figure 2

Box-Muller

method

LCG 2
$m = 16384$
$a = 9101$
$b = 1$
$x_0 = 0$

5000 points

The calculation of the trignometric functions used in the Box-Muller method can be substituded by a rejection method . In this way the following method [32] is a modification of the Box-Muller method.

Method 2: *The polar method transforms two independent, (0,1)-uniformly distributed random variables U_1, U_2 into the two independent, normally distributed random variables Y_1 and Y_2 (with mean 0 and standard deviation 1) if $(U_1 - \frac{1}{2})^2 + (U_2 - \frac{1}{2})^2 < 1$ by*

$$Y_1 = V_1 \cdot \sqrt{\frac{-2 \cdot \ln (V_1^2 + V_2^2)}{V_1^2 + V_2^2}} \qquad \text{and} \qquad Y_2 = V_2 \cdot \sqrt{\frac{-2 \cdot \ln (V_1^2 + V_2^2)}{V_1^2 + V_2^2}}$$

with $V_1 = 2 \cdot U_1 - 1$ and $V_2 = 2 \cdot U_2 - 1$.

Remark 2: *The polar method needs more uniform (pseudo-) random numbers than the Box-Muller method (average $4/\pi$). Only $\pi/4$ of the area of the unit square is transformed by the polar method. This may be the reason that the polar method seems to be a little bit more sensitive compared to the Box-Muller method (see figures 2, 3). But the sensitivity of the polar method is similar to the sensitivity of the Box-Muller method because this method is a two-to-two method too. For generators with large periods and good distributions in 2, 3, 4, . . . dimensions no great difference can be seen. (See also table 3 in section 3 of this paper).*

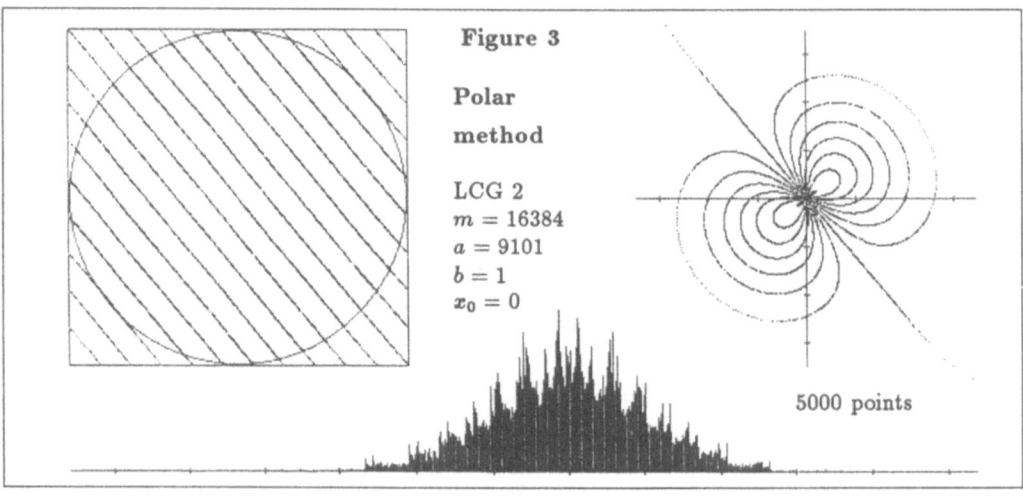

Figure 3

Polar method

LCG 2
$m = 16384$
$a = 9101$
$b = 1$
$x_0 = 0$

5000 points

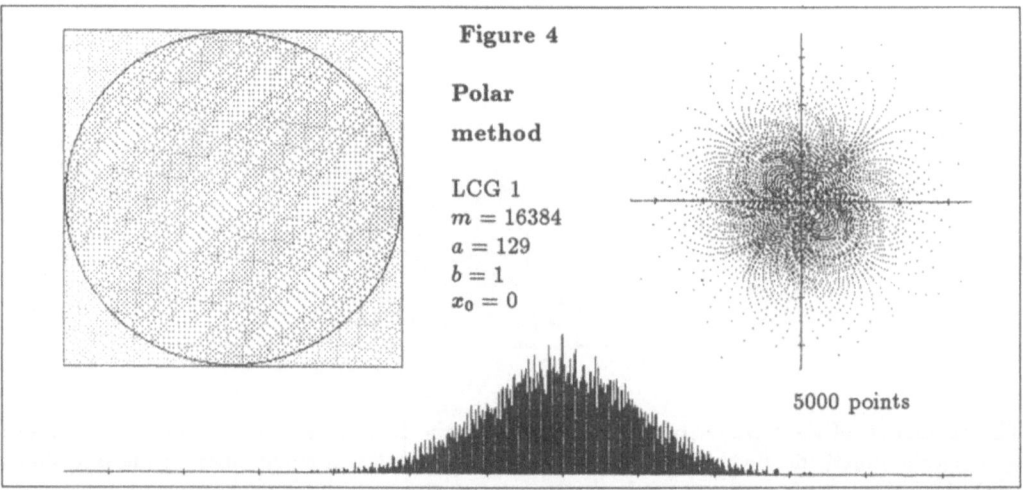

Figure 4

Polar method

LCG 1
$m = 16384$
$a = 129$
$b = 1$
$x_0 = 0$

5000 points

The Box-Muller method and the polar method are special transformation method to derive normally distributed (pseudo-) random numbers. There are general methods, like the following one, which can be used for various distributions [38], [28].

Method 3: *For an arbitrary density function f and $h(x) = k \cdot f(x)$, $x \in \mathbb{R}$, for a constant $k > 0$ and for two independent random variables U, V with (U, V) uniformly distributed on*

$$C = \left\{ (u, v) \in \mathbb{R}^2 : 0 < u \leq \sqrt{h\left(\frac{v}{u}\right)} \right\}$$

*the **ratio method** (or **ratio of uniforms method**) transforms the two random variables U, V by $Y = \frac{V}{U}$ into a random variable Y which has the density function f. Usually a rectangle is taken as a frame of C such that two independent, uniformly distributed random variables are needed in conjunction with a rejection method in order to get a uniformly distribution on C.*

In the case that f is the density of the standard normal distribution we have

$$C = \left\{ (u, v) \in \mathbb{R}^2 : 0 < u \leq 1, |v| \leq 2 \cdot u \cdot \sqrt{-\ln u} \right\} \subseteq [0, 1] \times \left[-\sqrt{\frac{2}{e}}, \sqrt{\frac{2}{e}} \right].$$

Remark 3: *The ratio method is a two-to-one method and it is much more sensitive than Box-Muller and polar method. The deficiencies of the uniform (pseudo-) random numbers in 2 dimensions can yield great deficiencies in one dimension as it can be seen in figure 5. For all points of the unit square which are lying on a half line from the point $(0, \frac{1}{2})$ the ratio method produces the same number $y = \frac{v}{u}$. There is a large gap from -2.54 to -1.78 in the histogram in figure 5.*

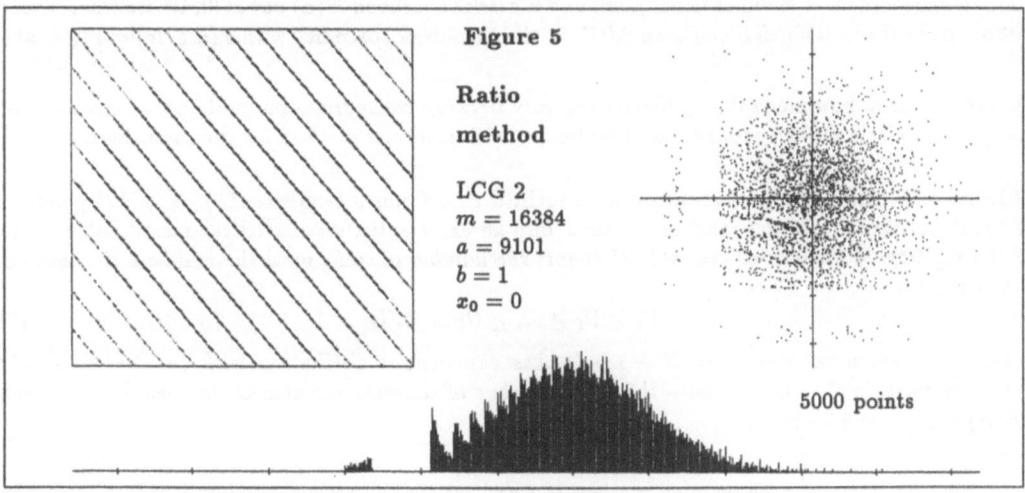

Figure 5

Ratio
method

LCG 2
$m = 16384$
$a = 9101$
$b = 1$
$x_0 = 0$

5000 points

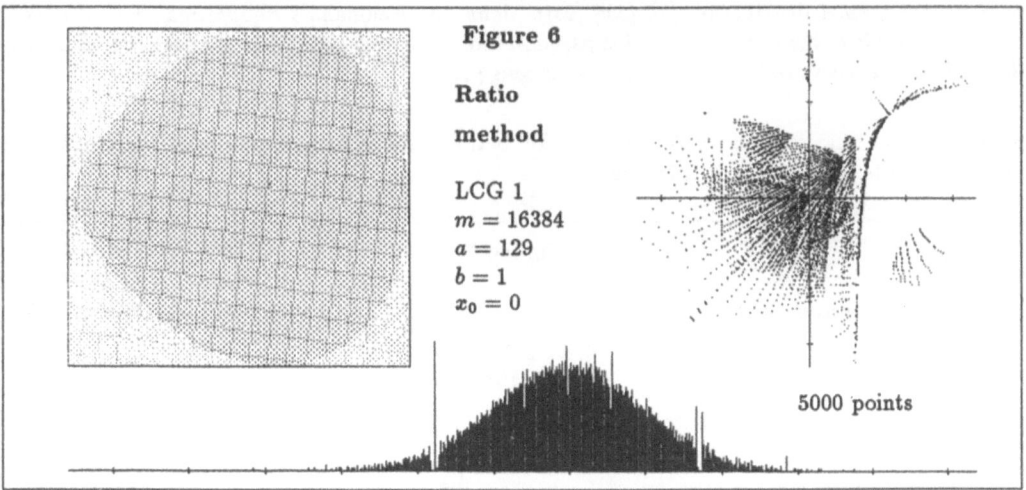

Figure 6

Ratio
method

LCG 1
$m = 16384$
$a = 129$
$b = 1$
$x_0 = 0$

5000 points

In the case of LCG 1 with a good two-dimensional distribution the deficiencies in the one-dimensional distribution of the transformed points are much smaller (see the histogram at ± 1.8 in figure 6). For better generators with large moduli these deficiencies are not to bee seen – the gaps are of the order of the two-dimensional resolution $1/\sqrt{m}$. The deficiencies of LCG 1 in $4, 6, \ldots$ dimensions can be seen in a bad two-dimensional distribution of the transformed points in figure 6. As LCG 2 has good distributions in $4, 6, \ldots$ dimensions the two-dimensional distribution of the transformed points in figure 5 seems to be the product distribution of the (bad) one-dimensional distributions.

The ratio method can also be used for other distributions. The sensitivity of the ratio method in the cases of Cauchy, Student-t and exponential distributions is similar to the case of normal distribution considerd above.

Method 4: *The inversion method transforms a (0,1)-uniformly distributed random variable U into a continuous random variable Y with the (invertible) distribution function F by Y = F⁻¹(U).*

Remark 4: *The inversion method can be used for various continuous and discrete distributions. We consider the sensitivity of this method in the case of a continuous distribution where we have a one-to-one method. A k-dimensional point and it's neighbourhood $N(\vec{u})$ out of $[0,1]^k$ are transformed into a point $\vec{y}$ and it's neighbourhood $M(\vec{y})$ in the Euclidean space $\mathbb{R}^k$ with $\mathrm{P}(\vec{u} \in N) = \mathrm{P}(\vec{y} \in M)$.*

In order to show how much the sensitivity can vary between transformation methods, we compare the inversion method for the exponential distribution with a special method for the same distribution.

Method 5: *The* **John von Neumann's method** *transforms a sequence $(U_i)_{i\in\mathbb{N}_0}$ of independent, (0,1)-uniformly distributed random variables into an exponential distributed random variable in the following way. If the random variable N counts the number of trails until the numbers K is even in the equation*

$$U_0 \geq U_1 \geq \ldots \geq U_{K-1} < U_K \tag{2}$$

then the random variable $Y = N - 1 + U_0$ has exponential distibution with parameter 1. The random variable A, which counts the total number of uniform variates U_i for one Y, has mean $E(A) = \frac{e}{1-e^{-1}} \approx 4.30026$.

Remark 5: *John von Neumann's method is very fast if the uniform (pseudo-) random numbers can be generated quickly. (A modifikation of this method, called Forsythe's method, is very often used for the normal distribution [9], [20], [23]). John von Neumann's comparison method is very much susceptible to the deficiencies in the pseudo-random number generators because special higher-dimensional qualities are needed in the comparisons (2).*

Figure 7 shows that the sensitivity differs a lot between the inversion and John von Neumann's methods for the exponential distribution.

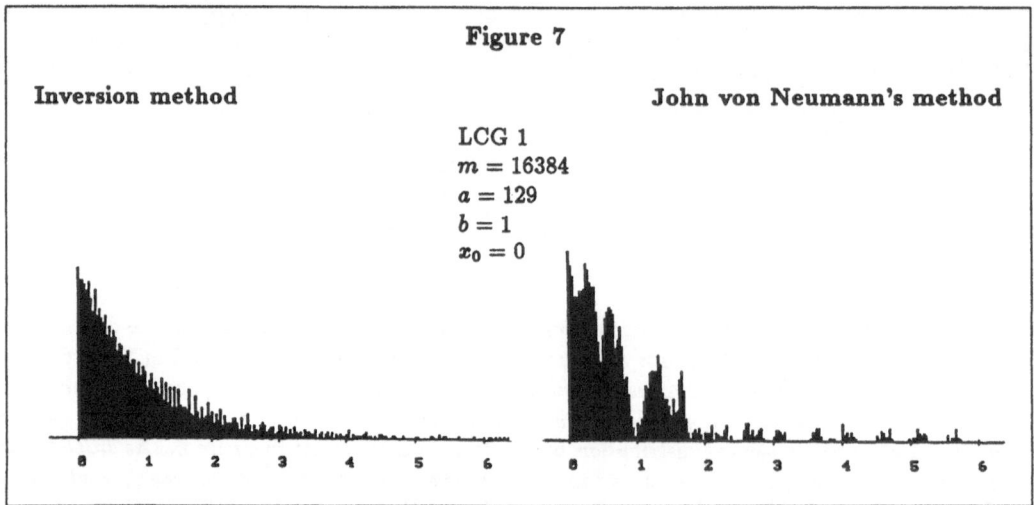

Figure 7

Inversion method   John von Neumann's method

LCG 1
$m = 16384$
$a = 129$
$b = 1$
$x_0 = 0$

Classification: We say a transformation method has the *sensitivity factor s* if the quality of the k-dimesional distribution of the transformed (nonuniform) points strongly depends on the quality of the $s \cdot k$-dimensional distribution of the uniform points used. The inversion method has sensitivity factor $s = 1$ because of the one-to-one relation. Nearly the same good behavior we have in the cases of the Box-Muller and the polar methods with the two-to-two relations and sensitivity factors $s = 1$. The ratio method has sensitivity factor $s = 2$, the one-dimensional distribution of the transformed numbers corresponds to the two-dimensional distribution of the uniform numbers (and it's resolution). In the case of John von Neumann's method the one-dimensional distribution of the transformed numbers depends on the quality of the uniform numbers in several dimensions, so we have sensitivity factors $s = 1, 2, 3, 4, \ldots$ for John von Neumann's method.

3   Numerical results

In this part we examine various transformation methods for the normal distribution. In addition to the methods of section 2 we consider some other methods too. As a measure for the quality of the approximation of the normal distribution we compute the one-dimensional **discrepancy** with respect to the normal distribution.

$$D_N^1(F) := \sup_{-\infty \leq s \leq t \leq \infty} \left| \frac{\#\{y_i : s < y_i \leq t, \, i = 1, \ldots, N\}}{N} - \Big(F(t) - F(s)\Big) \right| \qquad (3)$$

where F is the distribution function of the normal distribution. This discrepancy is calculated for various transformation methods combined with linear congruential generators (1) with large periods as used in application. LCG I, LCG II and LCG III were suggested as good generators. The multiplier a of LCG IV is smaller than $\sqrt{m}$ as proposed in [16] for fast computation. In the case of LCG IV we have $q_n > 0.567$ for the Beyer quotients of the lattices in $n = 2, \ldots, 20$ dimensions.

Table 1				
LCG I:	$m = 2^{32}$,	$a = 663608941$,	$b = 0$	suggested by Dieter, Ahrens in [21]
LCG II:	$m = 2^{32}$,	$a = 1099087573$,	$b = 0$	suggested by Fishman in [22]
LCG III:	$m = 2^{28}$,	$a = 532393$,	$b = 1$	suggested by Afflerbach in [3]
LCG IV:	$m = 2^{28}$,	$a = 15653$,	$b = 1$	mentioned above according to [16]

For these linear congruential generators we use all numbers of the generators for the compuation of the discrepancy (3). LCG I and LCG II have periods $L = m/4 = 2^{30}$. LCG III and LCG IV have periods $L = m = 2^{28}$. As a high discrepancy means that the parent population of all pseudo-random numbers returned by this method does not approximate the normal distribution very well. It is clear that a good transformation method should have a low discrepancy (3) for $N = L$. The lowest possible discrepancy is $1/L$, which is achieved for the inversion method (see method 4) combined with a linear congruential generator with period $L = m$ or $L = m/4$, but as the inversion method is very slow for the normal distribution it is necessary to compute the discrepancy of different transformation methods, which were proposed because of their speed or simplicity.

Among the large number of transformation methods for the normal distribution that use the idea of decomposition we examined the five algorithms of Table 2. To compute the discrepancy (3) for these methods it is necessary to decide what is the sequence of all pseudo-random numbers returned by a method as this sequence depends on the seed of the linear congruential generator. But for each linear congruential generator of table 1 the sequence of the normal pseudo-random numbers of these methods ends up in one cycle that is independent of the seed. Numerical tests have shown that the expected number of the normal pseudo-random numbers returned by the methods of table 2 before the sequence enters the cycle is smaller than 6. Therefore we neglect the starting part and consider the cycle as the sequence of all numbers generated by a certain method.

As the resolution of the pseudo-random numbers is better in low dimensions and as the cycle contains more numbers if the expected number of uniform deviates required is low we can conjecture that the methods using fewer uniform deviates has a lower discrepancy (see [4]).

To compute a lower bound of the discrepancy (3) we devide the real axis into 10^6 subintervals of equal probabilities and count the number of pseudo-random numbers lying in each interval. (These subintervalls are transformed with the distribution function F of the normal distribution into 10^6 subintervals of equal length in [0,1]). Thus it is possible to compute a lower bound

$$\hat{D}_Z^1(F) := \max_{\substack{-\infty \leq s \leq t \leq \infty \\ F(s) \cdot 10^6, F(t) \cdot 10^6 \in \mathbb{N}}} \left| \frac{\#\{y_i : s < y_i \leq t, i = 1, \ldots, Z\}}{Z} - \left(F(t) - F(s) \right) \right| \qquad (4)$$

of the discrepancy (3) for the normal distribution where Z is the length of the cycle of normal pseudo-random numbers returned by the respective transformation method. Table 1 gives the lower bounds of $L \cdot D_Z$ for the different decomposition methods combined with the linear congruential generators of table 1. As an upper bound for $L \cdot D_Z$ we have $L \cdot \hat{D}_Z^1(F) + L \cdot 2 \cdot 10^{-6}$ which can be obtained by adding $2^{30} \cdot 2 \cdot 10^{-6} = 2147.5$ for the LCG I and LCG II and $2^{28} \cdot 2 \cdot 10^{-6} = 536.9$ for LCG III and LCG IV.

Table 2		uniform	$L \cdot \hat{D}_Z^1(F)$			
Transformation method	suggested by	numbers	LCG I	LCG II	LCG III	LCG IV
Sum of three uniforms	Marsaglia, Bray '64	2.92	32659	28838	17349	22641
Largest triangular	Kinderman, Ramage '76	2.16	16088	18972	11200	12391
Acceptance complement	Hörmann, Derflinger '90	1.48	6235	11618	6190	6947
Alias method, rejection	Ahrens, Dieter '89	1.06	16068	2069	17713	18336
Almost exact inversion	Marsaglia '84	1.05	2899	2273	1448	2613

The first conclusion we can draw from Table 2 is the fact that the discrepancy (4) is strongly influenced by the transformation method used but not very much by the linear congruential generators (all generators have good lattice structures). Secondly the values given in Table 2 support the considerations that decomposition methods requiring few uniform random numbers perform better with respect to the discrepancy. Only the alias method [11] of Ahrens and Dieter seems to be an exception but this is due to the fact that the alias method needs 32-bit pseudo-random numbers to produce a satisfying result. For linear congruential generators with period 2^{32} the discrepancy is very close to the discrepancy of Marsaglia's almost exact inversion method [35].

Table 3 shows the one-dimensional discrepancy of the methods considered in section 2. As these methods combined with the generators with even period length L generate two distinct cycles we take both cycles for the computation of $\hat{D}_Z^{(1)}(F)$. In addition, the ratio method can be looked at as rejection from a table mountain (see [12]).

Table 3	$L \cdot \hat{D}_Z^1(F)$			
Transformation method	LCG I	LCG II	LCG III	LCG IV
Box-Muller	786	935	467	2469
Polar	642	772	465	2606
Ratio of uniforms	20388	39186	16722	13274
Table mountain rejection	684	359	238	5873

The Box-Muller and polar method have very low discrepancies except for the case of LCG IV where the multiplier is very small compared with the modulus. In this case the two-dimensional lattice is almost parallel to the axes which implies a slightly higher discrepancy. A much higher discrepancy is obtained for the ratio method. Theoretical considerations show that the large gap in the distribution of the normal numbers (see remark 3 in section 2) is an intervall $(s, t]$ with $F(t) - F(s) \approx \frac{1}{2\sqrt{L}}$ which yields $\sqrt{L}/2$ as a lower bound of $L \cdot D_Z^{(1)}(F)$. As the discrepancy is much lower for the rejection from a table mountain distribution generated by inversion (method 4 in table 3) the ratio method should be better replaced by this method. Only in the case of LCG IV the discrepancy is not very low because of the lattice parallel to the y-axis (see table 3).

4 Conclusion

The choice of a transformation method should not only be based on the computation time of the method as it is shown that the sensitivity can vary a lot between the transformation methods. Generators with good approximations of the n-dimensional distribution ($n = 1, 2, 3, \ldots, 15, \ldots, 20, \ldots$) should be used. Multiple recursive generators [26] or matrix generators [24], [25], [39], [41] with good lattice structures [6] could be used for better resolutions in higher dimensions in order to have good results for all methods.

References

[1] Afflerbach, L.: *Lineare Kongruenz-Generatoren zur Erzeugung von Pseudo-Zufallszahlen und ihre Gitterstruktur*. Dissertation, FB Mathematik, Technische Hochschule Darmstadt (1983).

[2] Afflerbach, L.: *The sub-lattice structure of linear congruential random number generators*. Manuscripta Math. **55**, 455-465 (1986).

[3] Afflerbach, L.: *Criteria for the Assessment of Random Number Generators*. J. Comput. and Applied Math. **31**, 3–10 (1990).

[4] Afflerbach, L.: *Die Gütebewertung von Pseudo-Zufallszahlen-Generatoren aufgrund theoretischer Analysen und algorithmischer Berechnungen*. Habilitationsschrift, Technische Hochschule Darmstadt (1990).

[5] Afflerbach, L., Grothe, H.: *Calculation of Minkowski-Reduced Lattice Bases*. Computing **35**, 269-276 (1985).

[6] Afflerbach, L., Grothe, H.: *The lattice structure of pseudo-random vectors generated by matrix generators*. J. Comput. Appl. Math. **23**, 127-131 (1988).

[7] Afflerbach, L., Wenzel, K.: *Normal random numbers lying on spirals and clubs*. Statistical Papers **29**, 237-244 (1988).

[8] Ahrens, J.H., Dieter, U.: *Computer methods for sampling from exponential and normal distributions*. Comm. Assoc. Comput. Mach. **15**, 873-882 (1972).

[9] Ahrens, J.H., Dieter, U.: *Extensions of Forsythe's Method for Random Sampling from the Normal Distribution*. Math. Comp. **27**, 927-937 (1973).

[10] Ahrens, J.H., Dieter, U.: *Nonuniform Random Numbers*. Manuscript (1988).

[11] Ahrens, J.H., Dieter, U.: *An alias method for sampling from the normal distribution*. Computing **42**, 159-170 (1989).

[12] Ahrens, J.H., Dieter, U.: *A convenient sampling method with bounded computation times for Poisson distributions*. American Journal of Math. and Management Sciences 1989: 1-13 (1989).

[13] Beyer, W.A.: *Lattice Structure and Reduced Bases of Random Vectors Generated by Linear Recurrences*. In: S.K. Zaremba (Ed.): Applications of Number Theory to Numerical Analysis, 361-370 (1972).

[14] Beyer, W.A., Roof, R.B., Williamson, D.: *The Lattice Structure of Multiplikative Pseudo-Random Vectors*. Math. Comp. **25**, 345-360 (1971).

[15] Box, J.E.P., Muller, M.E.: *A Note on the Generation of Random Normal Deviates*. Ann. Math. Statist. **29**, 610-611 (1958).

[16] Bratley, P., Fox, L., Schrage, E.: *A Guide to Simulation*. Springer-Verlag: New York, Berlin, Heidelberg, Tokyo (1983).

[17] Chay, S.C., Fardo, R.D., Mazumdar, M.: *On Using the Box-Muller Transformation with Multiplicative Congruential Pseudo-random Number Generators*. Appl. Statist. **24**, 132-135 (1975).

[18] Devroye, L.: *Non-Uniform Random Variate Generation*. New York, Berlin, Heidelberg, Tokyo: Springer (1986).

[19] Dieter, U.: *Statistical Independence of Pseudo-Random Numbers Generated by the Linear Congruential Method*. In: S.K. Zaremba (Ed.): Applications of Number Theory to Numerical Analysis, 287-317 (1972).

[20] Dieter, U., Ahrens, J.H.: *A combinatorical method for the generation of normally distributed random numbers*. Computing **11**, 137-146 (1973).

[21] Dieter, U., Ahrens, J.H.: *Uniform Random Numbers*. Inst. f. Math. Stat., Technische Hochschule Graz (1974).

[22] Fishman G. S.:*Multiplicative congruential random number generators with modulus 2^β : An exhaustive analysis for $\beta = 32$ and a partial analysis for $\beta = 48$*. University of North Carolina, Technical Report No UNC/OR/TR-87/10 (1987).

[23] Forsythe, G.E.: *Von Neumann's method for random sampling from the normal and other distributions*. Math. Comp. **26**, 817-826 (1972).

[24] Grothe, H.: *Matrixgeneratoren zur Erzeugung gleichverteilter Zufallsvektoren*. in: L. Afflerbach, J. Lehn (Eds.): Kolloquium über Zufallszahlen und Simulationen Teubner, Stuttgart (1986).

[25] Grothe, H.: *Matrix generators for pseudo-random vector generation*. Statistical Papers **28**, 233-238 (1987).

[26] Grube, A.: *Mehrfach rekursiv erzeugte Zahlenfolgen*. Dissertation, Fakultät f. Math., Universität Karlsruhe, (1973).

[27] Hörmann W., Derflinger G.: *The ACR method for generating normal random variables*. OR Spektrum, 181-185 (1990).

[28] Kinderman, A.J., Monaham, J.F.: *Computer generation of random variables using the ratio of uniform deviates*. ACM Trans. Math. Soft. **3** 257-260 (1977).

[29] Kinderman, A.J., Ramage J. G.: *Computer generation of normal random variables*. Journal of the American Statistical Association **71**, 893-896 (1976).

[30] Knuth, D.E.: *The Art of Computer Programming*. Vol. II, 2nd ed. Addison-Wesley: Reading (Mass.), Menlo Park (Cal.), London, Amsterdam, Don Mills (Ont.), Sydney (1969, 1981).

[31] Lehmer, D.H.: *Mathematical Methods in Large-Scale Computing Units*. Ann. Comp. Lab. Havard Univ. **26**, 141-146 (1951).

[32] Marsaglia, G.: Boeing Scientific Res. Lab. report D1-82-0203 (1962).

[33] Marsaglia, G.: *Random Number Fall Mainly in the Planes*. Proc. Nat.Acad.Sci. **61**, 25-28 (1968).

[34] Marsaglia, G.: *The Structure of Linear Congruential Sequences*. In: S.K. Zaremba (Ed.): Applications of Number Theory to Numerical Analysis, 249-285 (1972).

[35] Marsaglia, G.: *The exact-approximation method for generating random variables in a computer*. Journal of the American Statistical Association **79**, 218-221 (1984).

[36] Marsaglia, G., Bray, T.A.: *A convenient method for generating normal variables*. SIAM Review **6**, 260-264 (1964).

[37] Neave, .R.: *On Using the Box-Muller Transformation with Multiplicative Congruential Pseudo-Random Number Generators*. Appl. Statist., **22**, 92-97 (1973).

[38] Neumann, J. von: *Various techniques used in connection with random digits*. NBS App. Math. Ser. **12**, 36-38 (1951).

[39] Niederreiter, H.: *A pseudorandom vector generator based on finite field arithmetic*. Math. Japonica **31**, 759-774 (1986).

[40] Ripley, B.D.: *Stochastic Simulation*. John Wiley & Sons, New York (1987).

[41] Tahmi, E.-H.A.D.E.: *Contribution aux Générateurs des Vecteurs Pseudo-Aléatoires*. Thèse, Université des sciences et de la technologie Houari Boumedienne, Algier (1982).

NONLINEAR METHODS FOR PSEUDORANDOM NUMBER AND VECTOR GENERATION

HARALD NIEDERREITER

Institute for Information Processing
Austrian Academy of Sciences
Sonnenfelsgasse 19
A-1010 Vienna, Austria
E-mail: nied@qiinfo.oeaw.ac.at

1. Introduction

The principal aim of pseudorandom number generation is to devise and analyze deterministic algorithms for generating sequences of numbers which simulate a sequence of i.i.d random variables with given distribution function. We shall deal here exclusively with pseudorandom numbers for the uniform distribution on the interval $[0, 1]$, i.e. with *uniform pseudorandom numbers*. We refer to Knuth [16], Niederreiter [18], Ripley [25], and to the recent survey by Niederreiter [24] for a general background on uniform pseudorandom number generation.

The classical method for generating uniform pseudorandom numbers is the *linear congruential method*, where a large integer m and integers a, b are chosen and a sequence $y_0, y_1, \ldots \in \mathbf{Z}$ with $0 \leq y_n < m$ is generated by the recursion

$$y_{n+1} \equiv ay_n + b \bmod m \quad \text{for} \quad n = 0, 1, \ldots. \tag{1}$$

Then pseudorandom numbers are obtained by the normalization $x_n = y_n/m \in [0, 1)$ for $n = 0, 1, \ldots$. The modulus m depends on machine capabilities and is usually taken to be a prime or a power of 2. The choice of b is not so critical; in fact, one may just as well take $b = 0$. The statistical properties of the pseudorandom numbers depend foremost on the choice of the multiplier a in relation to m, so that great care has to be expended on this point (see Fishman [13] for a recent discussion). However, it is a fact of life that the linear congruential method has certain inherent deficiencies which cannot be overcome even by the most judicious choice of the multiplier. One such deficiency is the coarse lattice structure that is formed by the points $\mathbf{y}_n = (y_n, y_{n+1}, \ldots, y_{n+s-1}) \in \mathbf{Z}^s, n = 0, 1, \ldots$, in dimensions $s \geq 2$ (see [16], [25]). The underlying reason for this lattice structure is the simple linear character of the recursion (1).

In an attempt to get away from the troublesome lattice structure of linear congruential pseudorandom numbers, nonlinear methods for pseudorandom number generation have been introduced and studied in the last few years. This development is the theme of the present article. In Section 2 we will report on the current state of knowledge about general nonlinear congruential methods. A special nonlinear congruential method, the inversive congruential method, is particularly promising and will be reviewed in Section 3.

With the trend towards parallel computing there arises the need for pseudorandom vectors with good statistical properties. Such pseudorandom vectors can be used in parallelized simulation methods, multivariate computational statistics, and related areas. Up to now, nonlinear methods for pseudorandom vector generation have not been considered in the literature. We will propose such methods in Section 4.

We shall use the following conventions. The term "periodic" is always used in the sense of "purely periodic", i.e. a sequence of arbitrary elements $u_0, u_1, \ldots$ is *periodic* if there exists a positive integer T such that $u_{n+T} = u_n$ for all $n \geq 0$. The least value of T is the *least period length* of the sequence and denoted by $\mathrm{per}(u_n)$.

2. Nonlinear congruential methods

The general class of *nonlinear congruential generators* was introduced by Eichenauer, Grothe, and Lehn [2]. Let $p \geq 5$ be a large prime and generate $y_0, y_1, \ldots \in F_p = \{0, 1, \ldots, p-1\}$ by the recursion

$$y_{n+1} \equiv g(y_n) \bmod p \quad \text{for} \quad n = 0, 1, \ldots, \tag{2}$$

where g is an integer–valued function on F_p which is chosen in such a way that the sequence $y_0, y_1, \ldots$ is periodic with $\mathrm{per}(y_n) = p$. Then *nonlinear congruential pseudorandom numbers* are defined by $x_n = y_n/p \in [0, 1)$ for $n = 0, 1, \ldots$. In some cases it is also of interest to consider (2) with a composite modulus, e.g. in the *quadratic congruential method* where g is a quadratic polynomial (see Eichenauer and Lehn [4]) or in the *inversive congruential method* (see Section 3), but we will deal mainly with the case of a prime modulus.

An equivalent description of nonlinear congruential generators with prime modulus p is obtained by considering the mapping which takes $n \in F_p$ into $y_n \in F_p$. From now on we will often view F_p as the finite field with p elements, and then this mapping, like any self–mapping of F_p, can be represented by a uniquely determined polynomial f over F_p with $d := \deg(f) < p$. Since the y_n assume all values in F_p, the mapping in question is a bijection, and so we have $1 \leq d \leq p - 2$ (see [17, Ch. 7] for the facts about self–mappings of F_p that have been used here). The value of d is important in connection with the lattice test to be described below.

Definition 1. For an arbitrary sequence $y_0, y_1, \ldots \in F_p$ and a dimension $s \geq 1$ put

$$\mathbf{y}_n = (y_n, y_{n+1}, \ldots, y_{n+s-1}) \in F_p^s \quad \text{for} \quad n = 0, 1, \ldots.$$

Then the sequence $y_0, y_1, \ldots$ passes the *s–dimensional lattice test* if the vectors $\mathbf{y}_n - \mathbf{y}_0, n = 0, 1, \ldots,$ span the vector space F_p^s.

For the sake of comparison we first consider the s-dimensional lattice test for a linear congruential generator $y_0, y_1, \ldots$ obtained from (1) with a prime modulus p. For fixed $n \geq 0$ we get by induction on i that

$$y_{n+i} - y_i \equiv a^i (y_n - y_0) \bmod p \quad \text{for} \quad i = 0, 1, \ldots.$$

Therefore

$$\mathbf{y}_n - \mathbf{y}_0 = (y_n - y_0)(1, a, a^2, \ldots, a^{s-1}) \quad \text{for} \quad n = 0, 1, \ldots$$

in the vector space F_p^s. This shows that the vectors $\mathbf{y}_n - \mathbf{y}_0, n = 0, 1, \ldots,$ span a linear subspace of F_p^s of dimension at most 1. Consequently, a linear congruential generator with prime modulus can pass the s-dimensional lattice test only for $s = 1$. This is another instance of the weakness of linear congruential generators with regard to lattice structure.

The situation changes drastically when we turn to nonlinear congruential generators. The theorem below was shown by Eichenauer, Grothe, and Lehn [2] and a simpler proof was given by Niederreiter [19].

Theorem 1. *A nonlinear congruential generator passes the s-dimensional lattice test exactly for all $s \le d$.*

Thus, nonlinear congruential generators yielding a large value of d are preferable with respect to the lattice test. However, a word of warning is in order here: excellent behavior under the lattice test does not necessarily imply good statistical properties. For instance, a construction of Eichenauer and Niederreiter [6] shows that for any prime $p \ge 5$ there exists a nonlinear congruential generator mod p with very bad statistical properties, but which nevertheless achieves the maximal value $p - 2$ of d. We note that a special study of the lattice structure of quadratic congruential generators was carried out by Eichenauer and Lehn [4].

For the theoretical assessment of the statistical properties of uniform pseudorandom numbers we use the concept of discrepancy which is defined as follows.

Definition 2. For any N points $\mathbf{w}_0, \mathbf{w}_1, \ldots, \mathbf{w}_{N-1}$ in the s-dimensional interval $[0,1]^s, s \ge 1$, their *discrepancy* is defined by

$$D_N(\mathbf{w}_0, \mathbf{w}_1, \ldots, \mathbf{w}_{N-1}) = \sup_J \left| \frac{A(J; N)}{N} - V(J) \right|,$$

where the supremum is extended over all subintervals J of $[0,1]^s$, $A(J; N)$ is the number of $n, 0 \le n \le N - 1$, with $\mathbf{w}_n \in J$, and $V(J)$ denotes the volume of J.

Now let $x_0, x_1, \ldots$ be an arbitrary sequence of uniform pseudorandom numbers in $[0,1]$. For a given $s \ge 1$ we consider the sequence of s-dimensional points

$$\mathbf{x}_n = (x_n, x_{n+1}, \ldots, x_{n+s-1}) \in [0,1]^s \quad \text{for} \quad n = 0, 1, \ldots$$

and we put

$$D_N^{(s)} = D_N(\mathbf{x}_0, \mathbf{x}_1, \ldots, \mathbf{x}_{N-1})$$

for the discrepancy of the first N terms of this sequence. If $D_N^{(s)}$ is small for large N, then the sequence $x_0, x_1, \ldots$ passes the s-*dimensional serial test*; for $s = 1$ this is also called the *uniformity test*. If $x_0, x_1, \ldots$ is periodic, then it suffices to consider $N \le \text{per}(x_n)$.

For nonlinear congruential pseudorandom numbers it is easily seen that $D_p^{(1)} = 1/p$. Further information on the serial test is provided by the following result of Niederreiter [20] in which the number d from above again plays a role.

Theorem 2. *For nonlinear congruential pseudorandom numbers we have*

$$D_p^{(s)} = O\left(dp^{-1/2}(\log p)^s\right) \quad \text{for } 2 \le s \le d,$$

$$D_N^{(s)} = O\left(dN^{-1}p^{1/2}(\log p)^{s+1}\right) \quad \text{for } 1 \le N < p \text{ and } 1 \le s \le d - 1.$$

Further results in [20] show that Theorem 2 is in general best possible, in the sense that $D_p^{(s)}$ can be of an order of magnitude at least $p^{-1/2}$ and that the bounds cannot (or need not) hold for dimensions s higher than those allowed in Theorem 2. A detailed analysis of the two–dimensional serial test for the quadratic congruential method with prime–power modulus was recently carried out by Eichenauer–Herrmann and Niederreiter [10].

3. The inversive congruential method

A special nonlinear congruential method with particularly attractive properties is obtained if the operation of multiplicative inversion mod p is used. For $c \in F_p$ define $\bar{c} \in F_p$ by $c\bar{c} \equiv 1 \bmod p$ if $c \neq 0$ and $\bar{c} = 0$ if $c = 0$. The *inversive congruential method* of Eichenauer and Lehn [3] works with a large prime modulus p and generates a sequence $y_0, y_1, \ldots \in F_p$ by the recursion

$$y_{n+1} \equiv -a\bar{y}_n + b \bmod p \quad \text{for} \quad n = 0, 1, \ldots, \tag{3}$$

where $a, b \in F_p$ are constants. A sufficient condition for $y_0, y_1, \ldots$ to be periodic with $\operatorname{per}(y_n) = p$ is that the polynomial $x^2 - bx + a$ is primitive over F_p. Recall that a monic polynomial $f(x)$ of degree $k \geq 1$ over an arbitrary finite field F_q is a *primitive polynomial* over F_q if $f(0) \neq 0$ and if the least positive integer e such that $f(x)$ divides $x^e - 1$ in $F_q[x]$ is $q^k - 1$ (compare with [17, Ch. 3]). We assume from now on that $a, b \in F_p$ have been chosen in such a way that $x^2 - bx + a$ is a primitive polynomial over F_p. A sequence of *inversive congruential pseudorandom numbers* is obtained by $x_n = y_n/p \in [0, 1)$ for $n = 0, 1, \ldots$.

Inversive congruential generators behave very well under the lattice test. The following result of Niederreiter [19] represents a slight improvement on an earlier theorem of Eichenauer, Grothe, and Lehn [2].

Theorem 3. *An inversive congruential generator with prime modulus p passes the s-dimensional lattice test for all $s \leq (p+1)/2$.*

Example. For the prime modulus $p = 2^{31} - 1$, possible values of a and b such that the polynomial $x^2 - bx + a$ is primitive over F_p are given by $a = 7, b = 2^{31} - 1427039054$ according to Grothe [14]. By Theorem 3 the corresponding inversive congruential generator passes the s-dimensional lattice test for all $s \leq 2^{30}$.

The following strong nonlinearity property of inversive congruential generators mod p was shown by Eichenauer–Herrmann [7]. Here we view F_p^s as a finite affine space and consider the points $\mathbf{y}_n$ in Definition 1.

Theorem 4. *For $2 \leq s < p$ and every inversive congruential generator with prime modulus p, any hyperplane in F_p^s contains at most s of the points $\mathbf{y}_n$ with $0 \leq n \leq p - 1$ and $y_n \cdots y_{n+s-2} \neq 0$.*

Note that any s points in F_p^s define a hyperplane or a lower–dimensional linear manifold, and so Theorem 4 is optimal in the sense that there do exist hyperplanes in F_p^s which contain exactly s of the points $\mathbf{y}_n$ considered in Theorem 4.

As in the general nonlinear congruential method we have $D_p^{(1)} = 1/p$ for the one–dimensional discrepancy of a sequence of inversive congruential pseudorandom numbers. For higher dimensions we have the following result of Niederreiter [21].

Theorem 5. *For inversive congruential pseudorandom numbers with prime modulus p we have*

$$D_p^{(s)} = O\left(p^{-1/2}(\log p)^s\right) \quad for \quad 2 \le s < p.$$

The theorem below was proved in Niederreiter [22] and demonstrates that for a positive percentage of the parameters the bound in Theorem 5 is essentially best possible.

Theorem 6. *For a positive proportion of the possible parameter pairs (a, b) in the inversive congruential method with prime modulus p the generated pseudorandom numbers satisfy*

$$D_p^{(s)} \ge Cp^{-1/2} \quad for \ all \ s \ge 2,$$

where C is a positive absolute constant.

Theorems 5 and 6 show that in the inversive congruential method with prime modulus p the discrepancy $D_p^{(s)}$ has on the average an order of magnitude between $p^{-1/2}$ and $p^{-1/2}(\log p)^s$. We emphasize that it is in this range of magnitudes where one also finds the discrepancy of p independent and uniformly distributed random points from $[0, 1]^s$, according to the law of the iterated logarithm for discrepancies due to Kiefer [15].

It is an additional attractive feature of the inversive congruential method that once the basic condition on the parameters in (3) is satisfied, namely that $x^2 - bx + a$ be primitive over F_p, then the properties in Theorems 3,4, and 5 hold irrespective of the specific choice of parameters. Note that by [17, Theorems 3.5 and 3.16] the total number of primitive polynomials of degree 2 over F_p is given by $\phi(p^2 - 1)/2$, where ϕ is Euler's totient function. Thus, the inversive congruential method allows a wide choice of parameters, all of which lead to guaranteed and comparable statistical properties. This feature can be of great practical use in various parallelized simulation studies in which many parallel streams of pseudorandom numbers are needed. We refer to Niederreiter [24] for a comparison of the linear congruential and the inversive congruential method.

The inversive congruential method can also be used with a composite modulus m. Let G_m be the set of $c \in \mathbb{Z}$ with $1 \le c < m$ that are coprime to m. For $c \in G_m$ let $\bar{c}$ be the unique element of G_m with $c\bar{c} \equiv 1 \bmod m$. A sequence $y_0, y_1, \ldots \in G_m$ is generated by the recursion

$$y_{n+1} \equiv a\bar{y}_n + b \bmod m \quad \text{for} \quad n = 0, 1, \ldots. \tag{4}$$

The integers a, b have to be chosen in such a way that each y_n is guaranteed to be in G_m and that $y_0, y_1, \ldots$ is periodic with a large value of $\mathrm{per}(y_n)$. This was analyzed by Eichenauer, Lehn, and Topuzoğlu [5] in the case where $m \ge 8$ is a power of 2 and where a, y_0 are odd and b is even, and their result says that we get the maximal value $m/2$ of $\mathrm{per}(y_n)$ if and only if $a \equiv 1 \bmod 4$ and $b \equiv 2 \bmod 4$. A similar analysis was carried out by Eichenauer–Herrmann and Topuzoğlu [12] for arbitrary prime–power moduli. The lattice structure of inversive congruential generators with m a power of 2 was studied by Eichenauer-Herrmann et al. [9].

From a sequence $y_0, y_1, \ldots \in G_m$ generated by (4) we derive inversive congruential pseudorandom numbers by setting $x_n = y_n/m \in [0, 1)$ for $n = 0, 1, \ldots$. The results on the s-dimensional serial test for these pseudorandom numbers are not as complete as in the case of prime moduli. If m is a power of 2 and $a \equiv 1 \bmod 4, b \equiv 2 \bmod 4$, then it is easy to see that $D_{m/2}^{(1)} = 2/m$ and it was shown in Niederreiter [21] that $D_{m/2}^{(2)} = O\left(m^{-1/2}(\log m)^2\right)$, but results for higher dimensions are not yet available. On the other hand, for m a power of 2 an analog of Theorem 6 was established by Eichenauer-Herrmann and Niederreiter [11], in the sense that for a positive proportion of the possible parameter pairs (a, b) with $a \equiv 1 \bmod 4$ and $b \equiv 2 \bmod 4$ the discrepancy $D_{m/2}^{(s)}$ is at least of the order of magnitude $m^{-1/2}$ for all $s \ge 2$. Extensions of these results to arbitrary prime–power moduli have been obtained by Eichenauer-Herrmann [8].

4. Nonlinear methods for pseudorandom vector generation

The task in *uniform pseudorandom vector generation* is to simulate, for a given dimension $k \geq 2$, a sequence of independent and uniformly distributed random vector variables by a deterministic sequence of vectors (or points) from $[0,1]^k$. Possible applications of pseudorandom vectors have already been mentioned in Section 1. The first systematic studies of uniform pseudorandom vector generation have been devoted to *matrix generators*, which form the natural extension of linear congruential generators. A matrix generator uses the recursion (1), but with the y_n and b replaced by column vectors of dimension k and with a replaced by a square matrix of order k. Niederreiter [23] is the most recent paper on matrix generators and contains a brief history of this method.

Matrix generators share some of the deficiencies of linear congruential generators, for instance the coarse lattice structure (see Afflerbach and Grothe [1]). Therefore, it is of interest to consider also analogs of nonlinear congruential methods in the context of uniform pseudorandom vector generation. We propose such methods in the sequel.

The general *nonlinear method* for uniform pseudorandom vector generation proceeds as follows. Let p be a large prime and let F_q be the finite field with $q = p^k$ elements. Generate a sequence $\gamma_0, \gamma_1, \ldots \in F_q$ by the recursion

$$\gamma_{n+1} = g(\gamma_n) \quad \text{for} \quad n = 0, 1, \ldots, \tag{5}$$

where the self–mapping g of F_q is chosen in such a way that the sequence $\gamma_0, \gamma_1, \ldots$ is periodic with $\text{per}(\gamma_n) = q$. Now view F_q as a k-dimensional vector space over F_p and let $\{\beta_1, \ldots, \beta_k\}$ be a basis of F_q over F_p. Define the *trace*

$$\text{Tr}(\gamma) = \sum_{j=0}^{k-1} \gamma^{p^j} \quad \text{for} \quad \gamma \in F_q,$$

which is an F_p-linear transformation from F_q to F_p. Then from the sequence $\gamma_0, \gamma_1, \ldots$ above we derive a sequence of pseudorandom vectors by putting

$$\mathbf{u}_n = \frac{1}{p} \left(\text{Tr}(\beta_1 \gamma_n), \ldots, \text{Tr}(\beta_k \gamma_n) \right) \in [0,1)^k \quad \text{for} \quad n = 0, 1, \ldots. \tag{6}$$

Theorem 7. *The sequence* $\mathbf{u}_0, \mathbf{u}_1, \ldots$ *defined by (6) is periodic with* $\text{per}(\mathbf{u}_n) = p^k$. *Over the full period this sequence runs exactly through all the* p^k *points in the set* $p^{-1}\mathbf{Z}^k \cap [0,1)^k$.

Proof. Since $\gamma_0, \gamma_1, \ldots$ is periodic with $\text{per}(\gamma_n) = q = p^k$, it is clear that $\mathbf{u}_0, \mathbf{u}_1, \ldots$ is periodic with period length q. To prove that q is the least period length, we show that $\mathbf{u}_0, \mathbf{u}_1, \ldots, \mathbf{u}_{q-1}$ are all distinct. Suppose we had $\mathbf{u}_m = \mathbf{u}_n$ for some m and n with $0 \leq m < n \leq q - 1$. Then $\text{Tr}(\beta_j \gamma_m) = \text{Tr}(\beta_j \gamma_n)$ for $1 \leq j \leq k$. From the F_p-linearity of the trace and from the fact that $\{\beta_1, \ldots, \beta_k\}$ is a basis of F_q over F_p, it follows then that $\text{Tr}(\alpha \gamma_m) = \text{Tr}(\alpha \gamma_n)$ for all $\alpha \in F_q$. By [17, Theorem 2.24] this implies $\gamma_m = \gamma_n$, which is impossible. The second part of the theorem is obtained by noting that every $\mathbf{u}_n$ belongs to $p^{-1}\mathbf{Z}^k \cap [0,1)^k$ and that $\mathbf{u}_0, \mathbf{u}_1, \ldots, \mathbf{u}_{q-1}$ are all distinct as we have already proved. $\square$

A special nonlinear method is the *inversive method*. Here we choose $\alpha, \beta \in F_q$ such that $x^2 - \beta x + \alpha$ is a primitive polynomial over F_q (compare with Section 3). For $\gamma \in F_q$ define $\overline{\gamma} = \gamma^{-1}$ if $\gamma \neq 0$ and $\overline{\gamma} = 0$ if $\gamma = 0$. Generate $\gamma_0, \gamma_1, \ldots \in F_q$ by the recursion (5) with the function $g(\gamma) = -\alpha\overline{\gamma} + \beta$ for $\gamma \in F_q$. Then a sequence of pseudorandom vectors is derived by (6). To prove that this inversive method really falls into our category of nonlinear methods, it remains to show that the special function g has the property stated after (5). This is done in the following two lemmas.

Lemma 1. *If* $x^2 - \beta x + \alpha$ *is a primitive polynomial over* F_q, *then the sequence* $\sigma_0, \sigma_1, \ldots \in F_q$ *defined by* $\sigma_0 = 0, \sigma_1 = 1, \sigma_{n+2} = \beta \sigma_{n+1} - \alpha \sigma_n$ *for* $n = 0, 1, \ldots$ *is periodic with* $\mathrm{per}(\sigma_n) = q^2 - 1$ *and satisfies* $\sigma_n \neq 0$ *for* $1 \leq n \leq q$.

Proof. Since $\sigma_0, \sigma_1, \ldots$ is a maximal period sequence in F_q in the sense of [17, Definition 8.32], it is periodic with $\mathrm{per}(\sigma_n) = q^2 - 1$ by [17, Theorem 8.33]. Let η be a root of $x^2 - \beta x + \alpha$ in F_{q^2}. Then by [17, Theorem 8.24] there exists a $\theta \in F_{q^2}$ with $\theta \neq 0$ such that

$$\sigma_n = \theta \eta^n + \theta^q \eta^{qn} \quad \text{for} \quad n = 0, 1, \ldots . \tag{7}$$

Since $\sigma_0 = 0$, (7) shows that $\theta^{q-1} = -1$. Now suppose we had $\sigma_m = 0$ for some m with $1 \leq m \leq q$. Then (7) yields $\eta^{(q-1)m} = -\theta^{1-q} = 1$. Again by (7) we get then $\sigma_{n+(q-1)m} = \sigma_n$ for $n = 0, 1, \ldots$, thus $\mathrm{per}(\sigma_n) \leq (q-1)m \leq (q-1)q < q^2 - 1$, a contradiction. $\square$

Lemma 2. *If* $x^2 - \beta x + \alpha$ *is a primitive polynomial over* F_q, *then the sequence* $\gamma_0, \gamma_1, \ldots \in F_q$ *defined by* $\gamma_{n+1} = -\alpha \overline{\gamma}_n + \beta$ *for* $n = 0, 1, \ldots$ *is periodic with* $\mathrm{per}(\gamma_n) = q$.

Proof. Since $\alpha \neq 0$, the equation $\gamma_{n+1} = -\alpha \overline{\gamma}_n + \beta$ can be solved uniquely for γ_n if γ_{n+1} is given. Thus $\gamma_0, \gamma_1, \ldots$ is periodic with $\mathrm{per}(\gamma_n) \leq q$. Consider now the case where the initial value $\gamma_0 = 0$. If the sequence $\sigma_0, \sigma_1, \ldots$ is as in Lemma 1, then using $\sigma_n \neq 0$ for $1 \leq n \leq q$ one shows by induction on n that $\gamma_n = \sigma_{n+1}\sigma_n^{-1}$ for $1 \leq n \leq q$. Consequently we have $\gamma_n \neq 0$ for $1 \leq n \leq q - 1$, and so $\mathrm{per}(\gamma_n) \geq q$, hence $\mathrm{per}(\gamma_n) = q$. In particular $\{\gamma_0, \gamma_1, \ldots, \gamma_{q-1}\} = F_q$. If we have an arbitrary initial value γ_0, then the sequence $\gamma_0, \gamma_1, \ldots$ is a shifted version of the sequence with initial value 0, and so again $\mathrm{per}(\gamma_n) = q$. $\square$

It follows now from Theorem 7 that a sequence $u_0, u_1, \ldots$ generated by an inversive method is periodic with $\mathrm{per}(u_n) = p^k = q$ and that over the full period the sequence runs exactly through the set $p^{-1} \mathbf{Z}^k \cap [0, 1)^k$.

The serial test for sequences of k-dimensional pseudorandom vectors was introduced in Niederreiter [23]. We now apply this test to a sequence $u_0, u_1, \ldots$ generated by an inversive method. For a given dimension s define the points

$$\mathbf{v}_n = (u_n, u_{n+1}, \ldots, u_{n+s-1}) \in [0, 1)^{ks} \quad \text{for} \quad n = 0, 1, \ldots .$$

The s-*dimensional serial test* amounts to considering the discrepancy of long initial segments of the sequence $\mathbf{v}_0, \mathbf{v}_1, \ldots$. Since the standard discrepancy involves a large discretization error in this case (compare with [23, pp. 149–150]), it is more appropriate here to consider the *discrete discrepancy* $E_N^{(s)}$ of $\mathbf{v}_0, \mathbf{v}_1, \ldots, \mathbf{v}_{N-1}$ which is defined by extending the supremum in Definition 2 only over the subintervals J of $[0, 1]^{ks}$ of the form

$$J = \prod_{i=1}^{ks} \left[\frac{a_i}{p}, \frac{b_i}{p} \right)$$

with $a_i, b_i \in \mathbf{Z}$ and $0 \leq a_i < b_i \leq p$ for $1 \leq i \leq ks$. For $N = q$, i.e. for the full period, one can then show analogs of Theorems 5 and 6. The upper bound for the discrete discrepancy has the form $E_q^{(s)} = O\left(q^{-1/2}(\log p)^{ks}\right)$ for $2 \leq s < p$. There are $\phi(q^2 - 1)/2$ possible parameter pairs (α, β) in the inversive method, and for a positive proportion of those we have $E_q^{(s)} \geq Cq^{-1/2}$ for all $s \geq 2$, where $C > 0$ is an absolute constant. The proofs require deep results from the theory of finite fields and will be given in a later paper.

References

1. L. Afflerbach and H. Grothe: The lattice structure of pseudo–random vectors generated by matrix generators, *J. Comput. Appl. Math.* **23**, 127–131 (1988).

2. J. Eichenauer, H. Grothe, and J. Lehn: Marsaglia's lattice test and non–linear congruential pseudo random number generators, *Metrika* **35**, 241–250 (1988).

3. J. Eichenauer and J. Lehn: A non–linear congruential pseudo random number generator, *Statist. Papers* **27**, 315–326 (1986).

4. J. Eichenauer and J. Lehn: On the structure of quadratic congruential sequences, *Manuscripta Math.* **58**, 129–140 (1987).

5. J. Eichenauer, J. Lehn, and A. Topuzoğlu: A nonlinear congruential pseudorandom number generator with power of two modulus, *Math. Comp.* **51**, 757–759 (1988).

6. J. Eichenauer and H. Niederreiter: On Marsaglia's lattice test for pseudorandom numbers, *Manuscripta Math.* **62**, 245–248 (1988).

7. J. Eichenauer–Herrmann: Inversive congruential pseudorandom numbers avoid the planes, *Math. Comp.*, to appear.

8. J. Eichenauer–Herrmann: On the discrepancy of inversive congruential pseudorandom numbers with prime power modulus, preprint, Technische Hochschule Darmstadt, 1990.

9. J. Eichenauer–Herrmann, H. Grothe, H. Niederreiter, and A. Topuzoğlu: On the lattice structure of a nonlinear generator with modulus 2^α, *J. Comput. Appl. Math.* **31**, 81–85 (1990).

10. J. Eichenauer–Herrmann and H. Niederreiter: On the discrepancy of quadratic congruential pseudorandom numbers, *J. Comput. Appl. Math.*, to appear.

11. J. Eichenauer–Herrmann and H. Niederreiter: Lower bounds for the discrepancy of inversive congruential pseudorandom numbers with power of two modulus, preprint, Technische Hochschule Darmstadt, 1990.

12. J. Eichenauer–Herrmann and A. Topuzoğlu: On the period length of congruential pseudorandom number sequences generated by inversions, *J. Comput. Appl. Math.* **31**, 87–96 (1990).

13. G.S. Fishman: Multiplicative congruential random number generators with modulus 2^β: An exhaustive analysis for $\beta = 32$ and a partial analysis for $\beta = 48$, *Math. Comp.* **54**, 331–344 (1990).

14. H. Grothe: Matrixgeneratoren zur Erzeugung gleichverteilter Pseudozufallsvektoren, Dissertation, Technische Hochschule Darmstadt, 1988.

15. J. Kiefer: On large deviations of the empiric d.f. of vector chance variables and a law of the iterated logarithm, *Pacific J. Math.* **11**, 649–660 (1961).

16. D.E. Knuth: *The Art of Computer Programming*, Vol. 2: *Seminumerical Algorithms*, 2nd ed., Addison–Wesley, Reading, Mass., 1981.

17. R. Lidl and H. Niederreiter: *Finite Fields*, Addison–Wesley, Reading, Mass., 1983.

18. H. Niederreiter: Quasi–Monte Carlo methods and pseudo–random numbers, *Bull. Amer. Math. Soc.* **84**, 957–1041 (1978).

19. H. Niederreiter: Remarks on nonlinear congruential pseudorandom numbers, *Metrika* **35**, 321–328 (1988).

20. H. Niederreiter: Statistical independence of nonlinear congruential pseudorandom numbers, *Monatsh. Math.* **106**, 149–159 (1988).

21. H. Niederreiter: The serial test for congruential pseudorandom numbers generated by inversions, *Math. Comp.* **52**, 135–144 (1989).

22. H. Niederreiter: Lower bounds for the discrepancy of inversive congruential pseudorandom numbers, *Math. Comp.* **55**, 277–287 (1990).

23. H. Niederreiter: Statistical independence properties of pseudorandom vectors produced by matrix generators, *J. Comput. Appl. Math.* **31**, 139–151 (1990).

24. H. Niederreiter: Recent trends in random number and random vector generation, *Ann. Operations Research*, to appear.

25. B.D. Ripley: *Stochastic Simulation*, Wiley, New York, 1987.

SAMPLING FROM DISCRETE AND CONTINUOUS DISTRIBUTIONS WITH C–RAND

Ernst Stadlober   Ralf Kremer

Institut für Statistik, Techn. Universität Graz
Lessingstraße 27, A–8010 Graz, Austria
e–mail: statistik@rech.tu–graz.ada.at

Abstract

C–RAND is a system of Turbo–C routines and functions intended for use on microcomputers. It contains up–to–date random number generators for more than thirty univariate distributions. For some important distributions the user has the choice between extremely fast but rather complicated methods and somewhat slower but also much simpler procedures. Menu driven demo programs allow to test and analyze the generators with regard to speed and quality of the output.

1. Introduction

The purpose of our package is to provide random number generators for classical distributions as well as for nonstandard models which are of growing importance in simulation. However, most simulation packages may not be able to handle some of the more exotic models. Therefore the researcher my need some routines of our collection to perform his own specific simulation problem. For this he should be prepared at least to link the C–programs to his own software. The C–functions written in Turbo C 2.0 are designed for use on an IBM PC/AT or equivalent microcomputer with an 80287 coprocessor. The source codes of each generator are rather self-documented. They contain comments about references, the purpose of the specific algorithm, the meaning of input and output parameters and statements about subprograms needed. Detailed inline remarks should help to look through the programs if necessary. We decided also to incorporate only C–routines for the uniform generators in order to maintain the portability of the functions. Hence it should be no problem to transfer the programs and run them on workstations or other computers.

For most generators demonstration programs are accessible which run at the moment on PC's with EGA graphics environment (monocolor or color). With this the user may study some performance measures of generators, he could also look at graphs of distributions with respect to changing shape parameters or he could observe whether the output of different random number sequences from the same generator behave like random samples.

2. Uniform Random Number Generators

Random numbers u_i from the (0,1) uniform distribution ($U(0,1)$) are generated by the *multiplicative congruential method*. They are based on sequences

$$z_0 \equiv 1 \,(\text{mod } 4), \quad z_{i+1} \equiv a\, z_i \,(\text{mod } 2^E), \quad a \equiv 5 \,(\text{mod } 8). \tag{1}$$

During one period of length $n = 2^{E-2}$ all integers of the form $0 < 4k + 1 < m = 2^E$ are generated as z_i. The derived quotients $u_i = z_i/m$ are the required uniform random numbers in (0,1).

It is well known (see Ripley, 1987) that an adequate generator should have a long period and should produce random numbers u_i in such a way that the k–tuples $(u_i, \ldots, u_{i+k-1})$ are very close to the uniform distribution in the hypercube $(0,1)^k$, i.e. the generated k–tuples should fill a lattice in $(0,1)^k$. In C–RAND two generators — written in C as mentioned in the introduction — are implemented: the simple generator drand with period 2^{30} and the more complicated generator krand with longer period 2^{52}. Their special choices for multipliers a and moduls m are as follows.

$$\text{drand}: \quad a = 663608941, m = 2^{32}; \qquad \text{krand}: \quad a = 2783377641436325, m = 2^{54}.$$

The multipliers are close to $\theta\,n = \theta\,2^{E-2}$ where θ is the golden section number $\theta = \frac{1}{2}(\sqrt{5} - 1)$. These values are motivated by the theory of the distribution of pairs (u_i, u_{i+1}) in $(0,1)^2$ (see Dieter, 1971). The lattice structure up to 8 dimensions induced by these multipliers has been assessed by the spectral test of Coveyou and MacPherson (1967), and the results indicate that our choices behave 'well' also in higher dimensions. Our code of drand on an IBM PC is:

```
double drand(prand)
unsigned long *prand;
{
union ur { double d; int i[4];} x;
x.d = *prand *=0X278DDE6DL;
x.i[3] -=0X0200;
return(x.d);
}
```

*prand must be initialized in the calling program to some *prand = 4*k+1; ($k = 0, 1, \ldots$). Then the generator may be called by the C–statement u = drand(prand);. Standard initialization with *prand = 1; should result in the following sequence of random numbers: $u_1 = .15450849686749$, $u_2 = .98173871100880$, $u_3 = .35125617426820$, $u_4 = .82583231129684$, $u_5 = .54327537049538, \ldots$, $u_{20} = .46241982677020, \ldots, u_{50} = .92243475676514, \ldots, u_{100} = .32570007839240, \ldots$.

These values may be helpful to the user for checking the output of the generator on his own machine. All algorithms for generating nonuniform variates in the package use this basic uniform generator drand as stated above. But it could be easily replaced by krand if a longer period is preferred for some reason. Additionally the observed performance of the ready–made C–routines on the 80286 AT with 12 MHz 80287 coprocessor might be of interest.

drand occupies 313 bytes compiled code and needs $72\mu\text{sec}$ per sample.

krand occupies 570 bytes compiled code and needs $332\mu\text{sec}$ per sample.

3. Non–Uniform Random Number Generators

The main task of non–uniform random number generation is to convert the $(0,1)$–uniforms into variates that follow a given probability distribution $F(x)$. Many specific sampling algorithms for common distributions exist. They are based on theoretically exact transformation schemes: *inversion, rejection, rejection/composition, ratio of uniforms, acceptance/complement* and *distributional properties*. For a thorough description of the theory we refer to the books of Devroye (1986) and Bratley et al. (1987). For the development of our random number package we examined more than 80 algorithms and implemented the fastest generators found (see Kremer, 1989). We denote them as *efficient* procedures. In case of distributions with one or more shape parameters (e.g. gamma and Poisson (1 par.), beta and binomial (2 par.)) they are fast even when the parameters are frequently changing. However, several of these methods need long auxiliary tables and the programs are rather sophisticated. Hence we offer also so called *convenient* procedures for some distributions. These methods are based on a simple theory resulting in short programs and they still maintain their reasonable speed regardless of shifting parameter values.

3.1 Sampling from continuous distributions

The selected sampling methods for various continuous distributions without any shape parameter and with one or two shape parameters are listed in Table 1 below. It suffices to consider only algorithms for standardized distributions with location parameter $= 0$ and scale parameter $= 1$. A remarkable feature is that some of the methods are nothing but inversion by $X = F^{-1}(U)$ or $X = F^{-1}(1 - U)$ where $U \sim U(0, 1)$.

The very important case of the *exponential* distribution results in $X = F^{-1}(1 - U) = -\ln U$ which is now superior to different methods (Ahrens and Dieter, 1972, 1988), since in a PC environment with floating point coprocessor the evaluation of a logarithm is very cheap (~ 4 floating point additions).

For the *normal* distribution more than a dozen published methods have become fairly well-known. A *convenient* method is based on the transformation of Box and Muller (1958):

$$X = \sqrt{-2 \ln U} \cos(2\pi V), \quad Y = \sqrt{-2 \ln U} \sin(2\pi V), \tag{2}$$

which converts the two independent $(0,1)$–uniforms U and V into the two independent standard normal variates X and Y (sine–cosine method nsc). The fastest known normal generators rely on auxiliary tables containing tables of constants, which depend usually on the employed computer arithmetic (see Knuth, 1981). Recently, Ahrens and Dieter (1989) developed the *alias* method nal which is at least as fast as the older methods but simpler. Another advantage of nal is that the three tables of 128 one–byte integers needed do *not* depend on the precision of the arithmetic. However, the performance of the algorithm depends heavily on the programming language, since there is some bit or byte manipulation involved. Coding nal in C could be done very efficiently, because of its standard facilities for logical and shift operations. The execution times (compiled codes) of nal and nsc are $117\mu sec$ (1828 bytes) and $593\mu sec$ (624 bytes), respectively. nal needs 1.0618 uniforms per sample, but nsc requires exactly 1 uniform per normal variate.

Gamma distributions with shape parameters a received some attention in the literature (see Zechner, 1990, for an extensive comparison). The most efficient method gd ($a > 1$) is *acceptance/complement* with transformed normal variates (Ahrens and Dieter, 1982a), provided that a fast normal generator (in our case nal) is utilized. Our generator gds combines gd with the simple rejection method gs ($a \leq 1$) due to Ahrens and Dieter (1974). As *convenient* procedure rejection with log–logistic envelopes, denoted by gll, is employed (Cheng, 1977). The performance measures for gds and gll in case of $a \geq 1$ are listed below.

Algorithm	gds	gll
Execution times (μsec/sample)	560–1000	1300–2000
Set–up times (μsec/sample)	160 – 260	160 – 260
Compiled code (bytes)	5465	1113
No. of uniforms ($a = 1 \mid a \to \infty$)	1.97 | 1.57	2.94 | 2.26

Ratio of uniforms introduced by Kinderman and Monahan (1977) is the method of choice for *Chi(a)*, *Student–t(a)* and *generalized inverse Gaussian(λ, β)* distributions. The algorithms seem to be suitable combinations of convenience and speed. Good rejection algorithms exist for the *exponential–power(τ)*, *von Mises(k)* and *beta(a, b)* distributions. Generation from the *stable(a, d)* model is based on distributional properties. Sampling from the families of *Johnson $S_B, S_L, S_U(m, s)$* distributions is done by their definitions as transformed normal variates. For the members of the *Burr*–family and various other distributions closed form inversion of $F(x)$ is possible. Table 1 below gives an overview of generators for all continuous distributions covered by C–RAND.

Table 1. Generators for continuous distributions

Distribution	Algorithm	Method and Reference
No shape parameter		
Cauchy	cin	$X = \tan(\pi U)$
Exponential	eln	$= -\ln U$
Extreme valueI	ev_I	$= -\ln(-\ln U)$
Laplace (Double Exponential)	lapin	$= \pm \ln U$
Logistic	login	$= \ln(U/(1-U))$
Normal	nal	*Alias*, Ahrens/Dieter(1989)
	nsc	*sin/cos*, Box/Muller(1958)
Slash	sla	$X = Z/U,\ Z \sim N(0,1),\ U \sim U(0,1)$
Triangular	tra	$X = \pm(1 - \sqrt{U})$
One shape parameter		
BurrII(r), $r > 0$	bu_II	$X = -\ln(U^{-1/r} - 1)$, Burr(1942)
BurrVII(r), $r > 0$	bu_VII	$= \frac{1}{2}\ln((1+Y)/(1-Y)),\ Y = 2\,U^{1/r} - 1$
BurrVIII(r), $r > 0$	bu_VIII	$= \ln(\tan(\pi\,U^{1/r}/2))$
BurrX(r), $r > 0$	bu_X	$= \sqrt{-\ln(1 - U^{1/r})}$
Chi(a), $a > 1$	chru	*Ratio of Uniforms*, Monahan(1987)
Exp.–Power(τ), $\tau \geq 1$	epd	*Uniform–Exponential Rejection*, Devroye(1986)
Extreme valueII(a), $a > 0$	ev_II	$X = (-\ln U)^a$
Gamma(a), $0 < a \leq 1 / a > 1$	gds	*Rej./Acc.–Compl.*, Ahrens/Dieter(1974,1982a)
$a > 0$	gll	*Log-Logistic Rejection*, Cheng(1977)
Student–t(a), $a \geq 1$	trouo	*Ratio of Uniforms*, Kinderman/Monahan(1980)
von Mises(k), $k > 0$	mwc	*Wrapped Cauchy Rejection*, Best/Fisher(1979)
Weibull(c), $c > 0$	win	$X = (-\ln(1 - U))^{1/c}$
Two shape parameters		
Beta(a,b), $a,b > 0$	bbbc	*Log-Logistic Rejection*, Cheng(1978)
BurrIII(r,k), $r,k > 0$	bu_III	$X = Y^{-1/k},\ Y = U^{-1/r} - 1$
BurrIV(r,k), $r,k > 0$	bu_IV	$= k/(Y^k + 1)$
BurrV(r,k), $r,k > 0$	bu_V	$= \arctan(-\ln(Y/k))$
BurrVI(r,k), $r,k > 0$	bu_VI	$= \ln(Z + \sqrt{Z^2 + 1}),\ Z = -\frac{1}{r}\ln(Y/k)$
BurrIX(r,k), $r,k > 0$	bu_IX	$= \ln(W^{1/r} - 1),\ W = 1 + 2\,U/(k(1 - U))$
BurrXII(r,k), $r,k > 0$	bu_XII	$= Y^{1/k}$
G. I. Gauss(λ,β), $\lambda \geq 0, \beta > 0$	gigru	*Ratio of Uniforms*, Lehner(1989)
Johnson $S_B(m,s)$, $s > 0$	j_sb	$X = \exp(sZ + m)/(1 + \exp(sZ + m))$,
		$Z \sim N(0,1)$, Johnson(1949)
Johnson $S_L(m,s)$, $s > 0$	j_sl	$= \exp(sZ + m)$, (Lognormal)
Johnson $S_U(m,s)$, $s > 0$	j_su	$= \sinh(sZ + m)$
Lambda(λ_1,λ_2), $\lambda_1,\lambda_2 \geq 0$	lamin	$= U^{\lambda_1} - (1 - U)^{\lambda_2}$, Ramberg/Schmeiser(1974)
Stable(a,d), $0 < a \leq 2,\ |d| \leq 1$	stab	*Distributional Properties*, Chambers et al.(1976)

3.2 Sampling from discrete distributions

Efficient sampling methods have been developed for the classical *Poisson(μ)*, *binomial(n, p)* and *hypergeometric(N, M, n)* distributions. Poisson generator pd, due to Ahrens and Dieter (1982b), utilizes *acceptance–complement* with truncated normal variates whenever $\mu \geq 10$ and reverts to *table–aided inversion* otherwise. Binomial and hypergeometric distributions are handled by means of *composition–rejection* procedures (triangular–parallelogram–exponential method btpe and uniform–exponential method h2pe). Algorithms btpec (Kachitvichyanukul and Schmeiser 1988a, 1990) and h2pec (Kachitvichyanukul and Schmeiser 1985, 1988b) combine composition–rejection with *unstored inversion*, where inversion is performed as sequential search from the bottom using the recurrence relationships $p_k = g(p_{k-1})$.

Recently the *ratio of uniforms* method has been adapted to the distributions mentioned above (Poisson: Ahrens and Dieter, 1990, Stadlober, 1989a; binomial: Stadlober, 1989a, 1989b; hypergeometric: Stadlober, 1989a, 1990). It requires that the standardized histogram function $f(x) = p_k/p_m$, $k \leq x < k+1$, $p_m = \max_k p_k$, of the target discrete distribution is fitted under a (linearly transformed) *table mountain* $h(x) = \min(1, s^2/(x-a)^2)$ where a and s are suitably chosen parameters. With this the method can be stated as the following rejection procedure.

> Generate a pair (U, V) uniformly distributed over the rectangle $R = [0, 1] \times [-1, 1]$, set $X = sV/U + a$ and return $K = \lfloor X \rfloor$ as sample from $f(x)$ whenever $U^2 \leq f(X)$ is fulfilled. Otherwise reject X and try again.

This yields *convenient* algorithms with bounded computation times, i.e. the times do not grow with the parameters of the distributions. Again unstored or *simple inversion* is substituted in the utility routines pruec, bruec and hruec when it is faster than ratio of uniforms (see Table 2 below).

The *geometric* distribution with probability of success p and probability function $p_k = p(1 - p)^k$, $k = 0, 1, \ldots$ permit inversion by $K = \lfloor \ln U/(1-p) \rfloor$. Its generalization, the *negative binomial(r, p)* distribution is the discrete counterpart of the gamma distribution. Sampling is carried out by *simple inversion* whenever mean $\mu = r(1-p)/p \leq 10$. It is complemented for $\mu > 10$ by the *compound* method of Ahrens and Dieter (1974) which uses a relationship to gamma and Poisson variates. Generators for the *logarithmic(p)* distribution have been constructed by Kemp (1981), who suggested to use a combination of *inversion* ($p < .9$) and *distributional properties* ($p \geq .9$). Rare events in linguistics or insurance may be simulated by the two parameter *Zeta(ρ, τ)* distribution via a *rejection* scheme proposed by Dagpunar (1988).

A general discrete random number generator is based on the ingenious *alias* method of Walker (1977). It uses the fact that any discrete distribution with n mass points can be expressed as a mixture of $n-1$ equiprobable two point distributions. Two tables have to be constructed from the given probabilities p_k: the 'aliases' a_j (integers in $\{1, \ldots, n\}$) and the 'alias probabilities' q_j ($j = 1, \ldots, n-1$). The algorithm for this table set–up is of $O(n)$ time. The sampling procedure is very easy and uses exactly one uniform random number U.

> Generate $U \sim U(0, 1)$, transform it to $X = (n-1)U$ ($X \sim U(0, n-1)$), set $K = \lceil X \rceil$ ($K \sim U\{1, \ldots, n-1\}$) and $U = K - X$ ($U \sim U(0, 1)$ again). If $U \leq q_K$ return K ('criminal'), otherwise return a_K ('alias') as sample from p_k.

The method is extremely fast if the distribution stays fixed, but any change of parameters requires a new set–up procedure (new a_j and q_j).

Table 2. Generators for discrete distributions

Distribution	Algorithm	Method and Reference
One shape parameter		
Geometric(p), $0 < p < 1$	geo	$K = \lfloor \ln U/(1-p) \rfloor$
Logarithmic(p), $0 < p < 1$	lsk	*Inversion/Distr. Prop.*, Kemp(1981)
		$(p < .9)$ $(p \geq .9)$
Poisson(μ), $\mu > 0$	pd	*Table Inv./Acc.–Compl.*, Ahrens/Dieter(1982b)
		$(\mu < 10)$ $(\mu \geq 10)$
	pruec	*Inversion/Ratio of Unif.*, Stadlober(1989a)
		$(\mu < 5)$ $(\mu \geq 5)$
Two shape parameters		
Binomial(n,p), $n \geq 1, 0 < p < 1$,	btpec	*Inversion/Rej.–Comp.*, Kachitvichyanukul/
$\quad \ell = \min(np, n(1-p))$		$(\ell < 10)$ $(\ell \geq 10)$ $\quad$ Schmeiser(1988a,1990)
	bruec	*Inversion/Ratio of Unif.*, Stadlober(1989a,b)
		$(\ell < 5)$ $(\ell \geq 5)$
Neg. Bin.(r,p), $r > 0, 0 < p < 1$,	nbp	*Inversion/Compound*, Ahrens/Dieter(1974)
$\quad \mu = r(1-p)/p$		$(\mu < 10)$ $(\mu \geq 10)$
Zeta(ρ, τ), $\rho > 0, \tau \geq 0$	zet	*Rejection*, $\quad$ Dagpunar(1988)
Three shape parameters		
Hypergeometric(N, M, n),	h2pec	*Inversion/Rej.–Comp.*, Kachitvichyanukul/
$N, M, n \geq 1; n \leq N, M \leq N$,		$(\ell < 10)$ $(\ell \geq 10)$ $\quad$ Schmeiser(1985,1988b)
$m = \lfloor (n+1)(M+1)/(N+2) \rfloor$,	hruec	*Inversion/Ratio of Unif.*, Stadlober(1989a,1990)
$\ell = m - \max(0, n - N + M)$		$(\ell < 5)$ $(\ell \geq 5)$
Arbitrary distributions		
Given prob. $p_k, k = 1, \ldots, n$	gdal	*Alias Method*, $\quad$ Walker(1977)

4. An example of a demo program

Menu driven demo programs which run with EGA graphics card allow to visualize some features of the procedures implemented. As an example let us consider the flexible beta(a, b) distribution with generator bbbc of Cheng (1978). Starting the bbbc.exe program produces an image on the screen which is partitioned into seven windows. Two of these windows ('Beta Distribution', 'BBBC') are fixed, two windows allow for some input ('PAR...', 'Seed=...'), whereas in the three remaining windows information about the output is contained. The user can now easily select input windows in order to change some input parameters or he can request to open output windows for looking at the results. Output window 'Moments' delivers the mean number of uniforms needed for one sample, empirical and theoretical mean, variance, standard deviation, skewness and kurtosis of the random deviates. The execution times and set–up times (both in μsec per sample) appear in the window 'Timing'. In 'Graphics' the histogram of the generated random numbers is compared with the true density function marked by dots (see Figure 1). The 'true randomness' of the random samples may be studied by viewing histograms and moments of random number sequences with different seed values of the underlying uniform generator drand as demonstrated in Figure 2. Different shapes of the densities can be displayed by varying the shape parameters of the distribution ($a = 0.4, b = 0.6$: ' U–shaped'; $a = 0.8, b = 1.2$: 'J–shaped') as visualized in Figure 3.

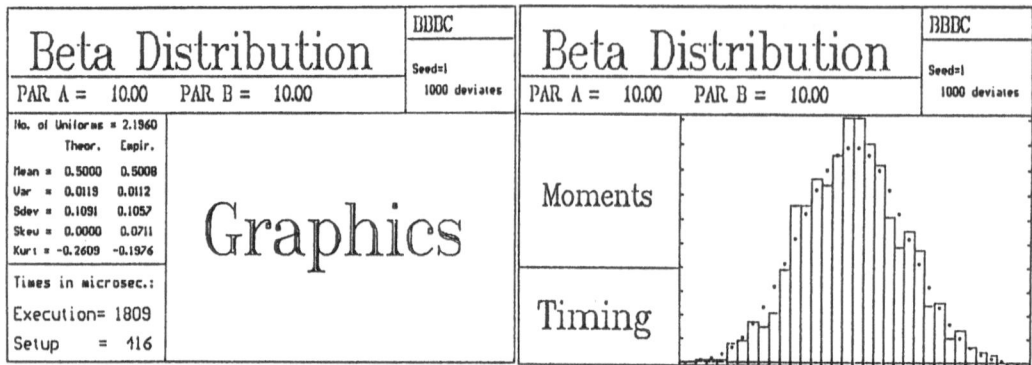

Figure 1: Moments, Timing and Graphics

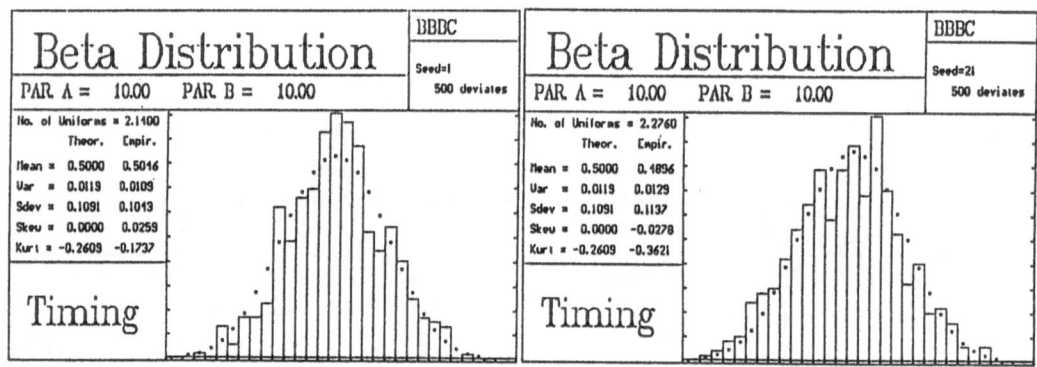

Figure 2: Effect of different seeds

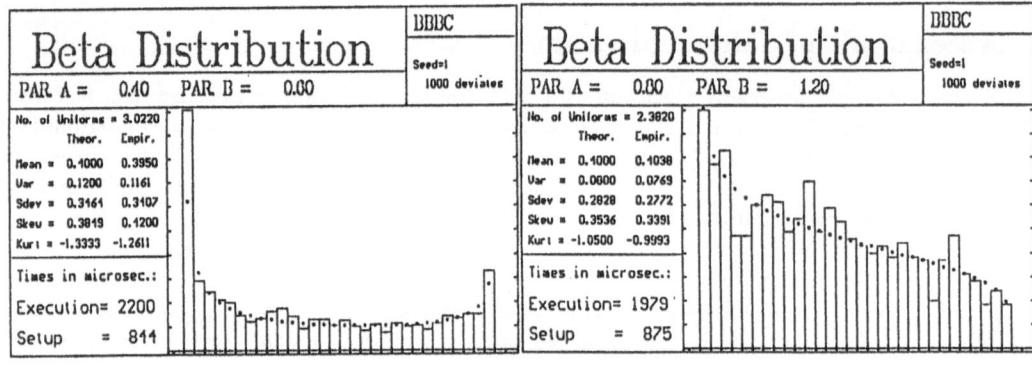

Figure 3: Varying shape parameters

Acknowledgement

We are grateful to Jo Ahrens for his valuable help in implementing the uniform generators drand, krand and his normal generator nal.

References

Ahrens, J.H. and Dieter, U. (1972): *Computer methods for sampling from the exponential and normal distributions*, Comm. ACM **15**, 873–882.

Ahrens, J.H. and Dieter, U. (1974): *Computer methods for sampling from gamma, beta, Poisson and binomial distributions*, Computing **12**, 223–246.

Ahrens, J.H. and Dieter, U. (1982a): *Generating gamma variates by a modified rejection technique*, Comm. ACM **25**, 47–54.

Ahrens, J.H. and Dieter, U. (1982b): *Computer generation of Poisson deviates from modified normal distributions*, ACM Trans. Math. Software 8, 163–179.

Ahrens, J.H. and Dieter, U. (1988): *Efficient table–free sampling methods for the exponential, Cauchy and normal distributions*, Comm. ACM **31**, 1330–1337.

Ahrens, J.H. and Dieter, U. (1989): *An alias method for sampling from the normal distribution*, Computing **42**, 159–170.

Ahrens, J.H. and Dieter, U. (1990): *A convenient sampling method with bounded computation times for Poisson distributions*, Amer. J. Math. Management Sci. **25**, in press.

Best, D.J. and Fisher, N.I. (1979): *Efficient simulation of the von Mises distribution*, Appl. Statist. **28**, 152–157.

Box, G.E.P. and Muller, M.E. (1958): *A note on the generation of random normal deviates*, Ann. Math. Statist. **29**, 610–611.

Bratley, P., Fox, B.L. and Schrage, L.E. (1987): *A Guide to Simulation*, 2nd Edition, Springer, New York.

Burr, I.W. (1942): *Cumulative frequency functions*, Ann. Math. Statist. **13**, 215–232.

Chambers, J.M., Mallows, C.L. and Stuck, B.W. (1976): *A method for simulating stable random variables*, J. Amer. Statist. Assoc. **71**, 340–344; (Correction: J. Amer. Statist. Assoc. **82**, (1987), 704).

Cheng, R.C.H. (1977): *The generation of gamma variables with non–integral shape parameter*, Appl. Statist. **26**, 71–75.

Cheng, R.C.H. (1978): *Generating beta variates with nonintegral shape parameters*, Comm. ACM **21**, 317–322.

Coveyou, R.R. and MacPherson, R.D. (1967): *Fourier analysis of uniform random number generators*, J. Assoc. Comput. Mach. **14**, 100–119.

Dagpunar, J. (1988): *Principles of Random Variate Generation*, Clarendon Press, Oxford.

Devroye, L. (1986): *Non–Uniform Random Variate Generation*, Springer, New York.

Dieter, U. (1971): *Pseudorandom numbers: the exact distribution of pairs*, Math. Comp. **25**, 855–883.

Johnson, N.L. (1949): *Systems of frequency curves generated by methods of translation*, Biometrika **36**, 149–176.

Kachitvichyanukul, V. and Schmeiser, B.W. (1985): *Computer generation of hypergeometric random variates*, J. Statist. Comput. Simulation **22**, 127–145.

Kachitvichyanukul, V. and Schmeiser, B.W. (1988a): *Binomial random variate generation*. Comm. ACM **31**, 216–222.

Kachitvichyanukul, V. and Schmeiser, B.W. (1988b): *ALGORITHM 668 H2PEC: Sampling from the hypergeometric distribution*, ACM Trans. Math. Software **14**, 397–398.

Kachitvichyanukul, V. and Schmeiser, B.W. (1990): *ALGORITHM BTPEC: Sampling from the binomial distribution*, ACM Trans. Math. Software **16**, to appear.

Kemp, A.W. (1981): *Efficient generation of logarithmically distributed pseudo–random variables*, Appl. Statist. **30**, 249–253.

Kinderman, A.J. and Monahan, J.F. (1977): *Computer generation of random variables using the ratio of uniform deviates*, ACM Trans. Math. Software **3**, 257-260.

Kinderman, A.J. and Monahan, J.F. (1980): *New methods for generating Student's t and gamma variables*, Computing **25**, 369–377.

Knuth, D.E. (1981): *The Art of Computer Programming, Vol. 2: Seminumerical Algorithms*, 2nd Edition, Addison Wesley, Reading.

Kremer, R. (1989): *C–RAND: Generatoren für nicht–gleichverteilte Zufallszahlen*, Diplomarbeit, 152 pp., Techn. Universität Graz.

Lehner, K. (1989): *Erzeugung von Zufallszahlen für zwei exotische stetige Verteilungen*, Diplomarbeit, 107 pp., Techn. Universität Graz.

Monahan, J.F. (1987): *An algorithm for generating chi random variables*, ACM Trans. Math. Software **13**, 168–172; (Correction: ACM Trans. Math. Software **14** (1988), 111).

Ramberg, J.S. and Schmeiser, B.W. (1974): *An approximate method for generating asymmetric random variables*, Comm. ACM **17**, 78–82.

Ripley, B.D. (1987): *Stochastic Simulation*, John Wiley, New York.

Stadlober, E. (1989a): *Sampling from Poisson, binomial and hypergeometric distributions: ratio of uniforms as a simple and fast alternative*, Math. Statist. Sektion **303**, 93 pp., Forschungsgesellschaft Joanneum Graz.

Stadlober, E. (1989b): *Ratio of uniforms as a convenient method for sampling from classical discrete distributions*, Proc. 1989 Winter Simulation Conf., Eds. E.A. MacNair et al., 484–489.

Stadlober, E. (1990): *The ratio of uniforms approach for generating discrete random variables*, J. Comput. Appl. Math. **31**, 181–189.

Walker, A.J. (1977): *An efficient method for generating discrete random variables with general distributions*, ACM Trans. Math. Software **3**, 253–256.

Zechner, H. (1990): *Erzeugung gammaverteilter Zufallszahlen mit allgemeinem Formparameter*, Grazer Math. Berichte **311**, 123 pp., Graz.

Lecture Notes in Economics and Mathematical Systems

For information about Vols. 1–210
please contact your bookseller or Springer-Verlag

Vol. 307: T.K. Dijkstra (Ed.), On Model Uncertainty and its Statistical Implications. VII, 138 pages. 1988.

Vol. 308: J.R. Daduna, A. Wren (Eds.), Computer-Aided Transit Scheduling. VIII, 339 pages. 1988.

Vol. 309: G. Ricci, K. Velupillai (Eds.), Growth Cycles and Multisectoral Economics: the Goodwin Tradition. III, 126 pages. 1988.

Vol. 310: J. Kacprzyk, M. Fedrizzi (Eds.), Combining Fuzzy Imprecision with Probabilistic Uncertainty in Decision Making. IX, 399 pages. 1988.

Vol. 311: R. Färe, Fundamentals of Production Theory. IX, 163 pages. 1988.

Vol. 312: J. Krishnakumar, Estimation of Simultaneous Equation Models with Error Components Structure. X, 357 pages. 1988.

Vol. 313: W. Jammernegg, Sequential Binary Investment Decisions. VI, 156 pages. 1988.

Vol. 314: R. Tietz, W. Albers, R. Selten (Eds.), Bounded Rational Behavior in Experimental Games and Markets. VI, 368 pages. 1988.

Vol. 315: I. Orishimo, G.J.D. Hewings, P. Nijkamp (Eds), Information Technology: Social and Spatial Perspectives. Proceedings 1986. VI, 268 pages. 1988.

Vol. 316: R.L. Basmann, D.J. Slottje, K. Hayes, J.D. Johnson, D.J. Molina, The Generalized Fechner-Thurstone Direct Utility Function and Some of its Uses. VIII, 159 pages. 1988.

Vol. 317: L. Bianco, A. La Bella (Eds.), Freight Transport Planning and Logistics. Proceedings, 1987. X, 568 pages. 1988.

Vol. 318: T. Doup, Simplicial Algorithms on the Simplotope. VIII, 262 pages. 1988.

Vol. 319: D.T. Luc, Theory of Vector Optimization. VIII, 173 pages. 1989.

Vol. 320: D. van der Wijst, Financial Structure in Small Business. VII, 181 pages. 1989.

Vol. 321: M. Di Matteo, R.M. Goodwin, A. Vercelli (Eds.), Technological and Social Factors in Long Term Fluctuations. Proceedings. IX, 442 pages. 1989.

Vol. 322: T. Kollintzas (Ed.), The Rational Expectations Equilibrium Inventory Model. XI, 269 pages. 1989.

Vol. 323: M.B.M. de Koster, Capacity Oriented Analysis and Design of Production Systems. XII, 245 pages. 1989.

Vol. 324: I.M. Bomze, B.M. Pötscher, Game Theoretical Foundations of Evolutionary Stability. VI, 145 pages. 1989.

Vol. 325: P. Ferri, E. Greenberg, The Labor Market and Business Cycle Theories. X, 183 pages. 1989.

Vol. 326: Ch. Sauer, Alternative Theories of Output, Unemployment, and Inflation in Germany: 1960–1985. XIII, 206 pages. 1989.

Vol. 327: M. Tawada, Production Structure and International Trade. V, 132 pages. 1989.

Vol. 328: W. Güth, B. Kalkofen, Unique Solutions for Strategic Games. VII, 200 pages. 1989.

Vol. 329: G. Tillmann, Equity, Incentives, and Taxation. VI, 132 pages. 1989.

Vol. 330: P.M. Kort, Optimal Dynamic Investment Policies of a Value Maximizing Firm. VII, 185 pages. 1989.

Vol. 331: A. Lewandowski, A.P. Wierzbicki (Eds.), Aspiration Based Decision Support Systems. X, 400 pages. 1989.

Vol. 332: T.R. Gulledge, Jr., L.A. Litteral (Eds.), Cost Analysis Applications of Economics and Operations Research. Proceedings. VII, 422 pages. 1989.

Vol. 333: N. Dellaert, Production to Order. VII, 158 pages. 1989.

Vol. 334: H.-W. Lorenz, Nonlinear Dynamical Economics and Chaotic Motion. XI, 248 pages. 1989.

Vol. 335: A.G. Lockett, G. Islei (Eds.), Improving Decision Making in Organisations. Proceedings. IX, 606 pages. 1989.

Vol. 336: T. Puu, Nonlinear Economic Dynamics. VII, 119 pages. 1989.

Vol. 337: A. Lewandowski, I. Stanchev (Eds.), Methodology and Software for Interactive Decision Support. VIII, 309 pages. 1989.

Vol. 338: J.K. Ho, R.P. Sundarraj, DECOMP: an Implementation of Dantzig-Wolfe Decomposition for Linear Programming. VI, 206 pages.

Vol. 339: J. Terceiro Lomba, Estimation of Dynamic Econometric Models with Errors in Variables. VIII, 116 pages. 1990.

Vol. 340: T. Vasko, R. Ayres, L. Fontvieille (Eds.), Life Cycles and Long Waves. XIV, 293 pages. 1990.

Vol. 341: G.R. Uhlich, Descriptive Theories of Bargaining. IX, 165 pages. 1990.

Vol. 342: K. Okuguchi, F. Szidarovszky, The Theory of Oligopoly with Multi-Product Firms. V, 167 pages. 1990.

Vol. 343: C. Chiarella, The Elements of a Nonlinear Theory of Economic Dynamics. IX, 149 pages. 1990.

Vol. 344: K. Neumann, Stochastic Project Networks. XI, 237 pages. 1990.

Vol. 345: A. Cambini, E. Castagnoli, L. Martein, P Mazzoleni, S. Schaible (Eds.), Generalized Convexity and Fractional Programming with Economic Applications. Proceedings, 1988. VII, 361 pages. 1990.

Vol. 346: R. von Randow (Ed.), Integer Programming and Related Areas. A Classified Bibliography 1984–1987. XIII, 514 pages. 1990.

Vol. 347: D. Ríos Insua, Sensitivity Analysis in Multi-objective Decision Making. XI, 193 pages. 1990.

Vol. 348: H. Störmer, Binary Functions and their Applications. VIII, 151 pages. 1990.

Vol. 349: G.A. Pfann, Dynamic Modelling of Stochastic Demand for Manufacturing Employment. VI, 158 pages. 1990.

Vol. 350: W.-B. Zhang, Economic Dynamics. X, 232 pages. 1990.

Vol. 351: A. Lewandowski, V. Volkovich (Eds.), Multiobjective Problems of Mathematical Programming. Proceedings, 1988. VII, 315 pages. 1991.

Vol. 352: O. van Hilten, Optimal Firm Behaviour in the Context of Technological Progress and a Business Cycle. XII, 229 pages. 1991.

Vol. 353: G. Riccil (Ed.), Declslon Processes In Economics. Proceedings, 1989. III, 209 pages 1991.

Vol. 354: M. Ivaldi, A Structural Analysis of Expectation Formation. XII, 230 pages. 1991.

Vol. 355: M. Salomon. Deterministic Lotsizing Models for Production Planning. VII, 158 pages. 1991.

Vol. 356: P. Korhonen, A. Lewandowski, J . Wallenius (Eds.), Multlple Crltena Decision Supporl. Proceedings, 1989. XII, 393 pages. 1991.

Vol. 358: P. Knottnerus, Linear Models with Correlaled Disturbances. VIII, 196 pages. 1991.

Vol. 359: E. de Jong, Exchange Rate Determination and Optimal Economlc Policy Under Various Exchange Rate Regimes. VII, 270 pages. 1991.